Pinpoint Math™

Student Booklet
Level E

Volume 5
Data, Geometry, and Measurement

Photo Credits

©iStock International Inc., cover.

Acknowledgements

Content Consultant:

Linda Proudfit, Ph.D.

After earning a B.A. and M.A in Mathematics from the University of Northern Iowa, Linda Proudfit taught junior- and senior-high mathematics in Iowa. Following this, she earned a Ph.D. in Mathematics Education from Indiana University. She currently is Coordinator of Elementary Education and Professor of Mathematics Education at Governors State University in University Park, IL.

Dr. Proudfit has made numerous presentations at professional meetings at the local, state, and national levels. Her main research interests are problem solving and algebraic thinking.

www.WrightGroup.com

Copyright © 2009 by Wright Group/McGraw-Hill.

All rights reserved. Except as permitted under the United States Copyright Act, no part of this publication may be reproduced or distributed in any form or by any means, or stored in a database or retrieval system, without the prior written permission from the publisher, unless otherwise indicated.

Printed in USA.

Send all inquiries to:
Wright Group/McGraw-Hill
P.O. Box 812960
Chicago, IL 60681

ISBN 978-1-40-4568044
MHID 1-40-4568042

2 3 4 5 6 7 8 9 10 RHR 13 12 11 10 09

Contents

Tutorial Chart viii

Volume 5: Data, Geometry, and Measurement

Topic 12 Graphing

Topic 12 Introduction. 1
Lesson 12-1 Compare Data with Graphs 2–4
Lesson 12-2 Record Data 5–7
Lesson 12-3 Mean, Median, and Mode 8–10
Lesson 12-4 Show Data in More than One Way 11–13
Lesson 12-5 Graph Ordered Pairs 14–16
Topic 12 Summary. 17
Topic 12 Mixed Review 18

Topic 13 Basic Geometric Figures

Topic 13 Introduction. 19
Lesson 13-1 Angles and Lines 20–22
Lesson 13-2 Types of Polygons. 23–25
Lesson 13-3 Triangles. 26–28
Lesson 13-4 Quadrilaterals 29–31
Lesson 13-5 Circles 32–34
Topic 13 Summary. 35
Topic 13 Mixed Review. 36

Topic 14 Measurement Conversion

Topic 14 Introduction 37
Lesson 14-1 U.S. Customary Units 38–40
Lesson 14-2 Basic Metric Prefixes 41–43
Lesson 14-3 Use the Metric System 44–46
Lesson 14-4 Factors in Unit Conversions 47–49
Lesson 14-5 Convert Units within a System 50–52
Topic 14 Summary 53
Topic 14 Mixed Review 54

Topic 15 Measure Geometric Figures

Topic 15 Introduction. 55
Lesson 15-1 Length 56–58
Lesson 15-2 Perimeter 59–61
Lesson 15-3 Area 62–64
Lesson 15-4 Area of Rectangles. 65–67
Lesson 15-5 Volume 68–70
Topic 15 Summary. 71
Topic 15 Mixed Review 72

Glossary. 73

Word Bank 76

Index 78

Objectives

Volume 5: Data, Geometry, and Measurement
Topic 12 Graphing

Lesson	Objective	Pages
Topic 12 Introduction	**12.1** Represent and compare data by using pictures, bar graphs, tally charts, and picture graphs. **12.4** Represent the same data in more than one way. **12.6** Use two-dimensional coordinate graphs to represent points and graph lines and simple figures. **12.7** Identify ordered pairs of data from a graph and interpret the meaning of the data in terms of the situation depicted by the graph.	1
Lesson 12-1 Compare Data with Graphs	**12.1** Represent and compare data by using pictures, bar graphs, tally charts, and picture graphs.	2–4
Lesson 12-2 Record Data	**12.2** Record numerical data in systematic ways, keeping track of what has been counted.	5–7
Lesson 12-3 Mean, Median, and Mode	**12.3** Find the mean, median, and mode of a set of data.	8–10
Lesson 12-4 Show Data in More than One Way	**12.4** Represent the same data in more than one way.	11–13
Lesson 12-5 Graph Ordered Pairs	**12.5** Use two-dimensional coordinate graphs to represent points and graph lines and simple figures.	14–16
Topic 12 Summary	Review skills related to graphing.	17
Topic 12 Mixed Review	Maintain concepts and skills.	18

Topic 13 Basic Geometric Figures

Lesson	Objective	Pages
Topic 13 Introduction	**13.1** Draw, measure, and classify different types of angles and lines. **13.2** Define polygon and classify the different types of polygons. **13.3** Explore, compare, and classify different types of triangles. **13.5** Explore circles and define their parts.	19
Lesson 13-1 Angles and Lines	**13.1** Draw, measure, and classify different types of angles and lines.	20–22
Lesson 13-2 Types of Polygons	**13.2** Define *polygon* and classify the different types of polygons.	23–25
Lesson 13-3 Triangles	**13.3** Explore, compare, and classify different types of triangles.	26–28
Lesson 13-4 Quadrilaterals	**13.4** Explore, compare, and classify different types of quadrilaterals.	29–31
Lesson 13-5 Circles	**13.5** Explore circles and define their parts.	32–34
Topic 13 Summary	Review basic geometric figures.	35
Topic 13 Mixed Review	Maintain concepts and skills.	36

Topic 14 Measurement Conversion

Lesson	Objective	Pages
Topic 14 Introduction	**14.5** Carry out simple unit conversions within a system of measurement.	37
Lesson 14-1 U.S. Customary Units	**14.1** Explore the basic units of measure in the United States.	38–40
Lesson 14-2 Basic Metric Prefixes	**14.2** Explore the basic metric prefixes and what they mean.	41–43
Lesson 14-3 Use the Metric System	**14.3** Explore the basic metric units and their relationships.	44–46
Lesson 14-4 Factors in Unit Conversions	**14.4** Express simple unit conversions in symbolic form.	47–49
Lesson 14-5 Convert Units within a System	**14.5** Carry out simple unit conversions within a system of measurement.	50–52
Topic 14 Summary	Review measurement conversions.	53
Topic 14 Mixed Review	Maintain concepts and skills.	54

Topic 15 Measure Geometric Figures

Lesson	Objective	Pages
Topic 15 Introduction	**15.1** Measure the length of an object to the nearest inch or centimeter. **15.2** Find the perimeter of a polygon with integer sides. **15.3** Estimate or determine the area of figures by covering them with squares. **15.4** Estimate or determine the volume of solid figures by counting the number of cubes that would fill them.	55
Lesson 15-1 Length	**15.1** Measure the length of an object to the nearest inch or centimeter.	56–58
Lesson 15-2 Perimeter	**15.2** Find the perimeter of a polygon with integer sides.	59–61
Lesson 15-3 Area	**15.3** Estimate or determine the area of figures by covering them with squares.	62–64
Lesson 15-4 Area of Rectangles	**15.4** Measure the area of rectangular shapes by using appropriate units.	65–67
Lesson 15-5 Volume	**15.5** Estimate or determine the volume of solid figures by counting the number of cubes that would fill them.	68–70
Topic 15 Summary	Review measuring geometric figures.	71
Topic 15 Mixed Review	Maintain concepts and skills.	72

Tutorial Guide

Each of the standards listed below has at least one animated tutorial for students to use with the lesson that matches the objective. If you are using the electronic components of *Pinpoint Math*, you will find a complete listing of Tutorial codes and titles when you access them either online or via CD-ROM.

Level E

Standards by topic	Tutorial codes
Volume 5 Data, Geometry, and Measurement	
Topic 12 Meaning of Fractions	
12.1 Represent and compare data by using picture, bar graphs, tally charts, and picture graphs.	12a Using a Tally Chart to Make a Bar Graph
12.2 Record numerical data in systematic ways, keeping track of what has been counted.	12b Using a Data Table to Make a Bar Graph
12.3 Find the mean, median, and mode of a set of data.	12c Find the Mean, Median, and Mode of a Data Set
12.4 Represent the same data in more than one way.	12a Using a Tally Chart to Make a Bar Graph
12.4 Represent the same data in more than one way.	12b Using a Data Table to Make a Bar Graph
12.5 Use two-dimensional coordinate graphs to represent points and graph lines and simple figures.	12d Graphing Ordered Pairs
Topic 13 Basic Geometric Figures	
13.1 Draw, measure, and classify different types of angles and lines.	13a Identifying and Drawing Parallel Lines
13.1 Draw, measure, and classify different types of angles and lines.	13b Identifying and Drawing Perpendicular Lines
13.2 Define *polygon* and classify the different types of polygons.	13e Classifying Polygons
13.3 Explore, compare, and classify different types of triangles.	13f Sorting and Classifying Triangles
13.4 Explore, compare, and classify different types of quadrilaterals.	13h Finding Angles Measures in Quadrilaterals
13.5 Explore circles and define their parts.	13i Finding the Circumference of a Circle, Given a Radius
Topic 14 Measurement Conversion	
14.1 Explore the basic units of measure in the United States.	14a Converting Units of Capacity
14.2 Explore the basic metric prefixes and what they mean.	14b Using the Metric System to Measure Length
14.2 Explore the basic metric prefixes and what they mean.	14c Using the Metric System to Measure Mass
14.3 Explore the basic metric units and their relationships.	14b Using the Metric System to Measure Length
14.3 Explore the basic metric units and their relationships.	14c Using the Metric System to Measure Mass
14.4 Express simple unit conversions in symbolic form.	14d Converting Units of Time
14.5 Carry out simple unit conversions within a system of measurement.	14e Converting Units of Length
Topic 15 Measure Geometric Figures	
15.1 Measure the length of an object to the nearest inch or centimeter.	15a Measuring Length to the Nearest Unit
15.2 Find the perimeter of a polygon with integer sides.	15b Finding Perimeter
15.3 Estimate or determine the area of figures by covering them with squares.	15c Finding Area
15.5 Estimate or determine the volume of solid figures by counting the number of cubes that would fill them.	15d Finding Volume

Topic 12: Graphing

Topic Introduction

Complete with teacher help if needed.

1. **Books Read Last Month**

   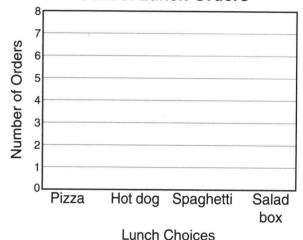

   a. Who read fewer than 4 books?

   _____

   b. Who read more books than Tiva?

   _____

   Objective 12.1: Represent and compare data by using pictures, bar graphs, tally charts, and picture graphs.

2. The following are the school lunch orders for Friday: pizza 8 orders, hot dog 7 orders, spaghetti 5 orders, and salad box 2 orders.

   Use the information above to complete the bar graph.

   School Lunch Orders

   Objective 12.4: Represent the same data in more than one way.

3. **Children in the Park at Noon**

   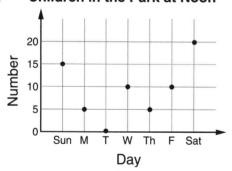

   When were there no children in the park?

   _____

   Objective 12.6: Use two-dimensional coordinate graphs to represent points and graph lines and simple figures.

4. **Jaycek's Shirt Factory**

   Each shirt needs 5 buttons. Mark points to show the buttons needed for 2, 4, and 6 shirts. Draw a line through the points. How many shirts can be made with 40 buttons?

   _____

   Objective 12.6: Use two-dimensional coordinate graphs to represent points and graph lines and simple figures.

| Lesson 12-1 | Compare Data with Graphs |

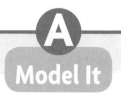

Words to Know The **range** is the difference between the greatest and least numbers in a set of data.

Activity 1

Use the graph. How many more students chose cats than birds?

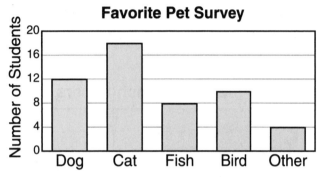

The bar for cats is 18 units tall. The bar for birds is 10 units. Subtract to compare the data.

$18 - 10 = 8$

There are 8 more students who prefer cats.

Practice 1

Use the graph. How many more students walk than ride their bikes?

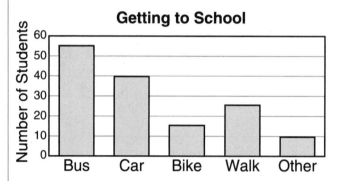

The bar for walking is _____ units tall.

The bar for biking is _____ units tall.

How many more walk? _____

Activity 2

Use the Favorite Pet graph.

What is the most popular? Cats

How many people prefer them? 18

What is the least popular? Other

How many people prefer them? 4

What is the range? $18 - 4 = 14$

Practice 2

Use the Getting to School graph.

What is the most popular? _____

How many people use it? _____

What is the least popular? _____

How many people use it? _____

What is the range? _____ − _____ = _____

On Your Own

Use the Favorite Pet graph. How many fewer people prefer fish than dogs?

_____

Write About It

Use the information in the Getting to School graph. Write two new comparisons.

Objective 12.1: Represent and compare data by using pictures, bar graphs, tally charts, and picture graphs.

Lesson 12-1 — Compare Data with Graphs

Understand It

Words to Know A **pictograph** is a graph that uses symbols or simple pictures to represent quantities.
The **key** on a pictograph shows the value of each symbol.

Example 1

How many more customers were there in June than in April?

Flower Shop Customers

April	👤 👤 👤 👤 👤 👤
May	👤 👤 👤 ᒧ
June	👤 👤 👤 👤 👤 👤 👤 ᒧ

👤 = 10 customers

For June and April, multiply the number of symbols by 10. Subtract to compare.

June: $7\frac{1}{2} \times 10 = 75$ April: $6 \times 10 = 60$

June − April = 75 − 60 = 15

June had 15 more customers than April.

Practice 1

How many more hours are there for Grade 6 than for Grade 7?

Volunteer Hours

Grade 4	☆ ☆ ☆ ☆
Grade 5	☆ ☆ ᒧ
Grade 6	☆ ☆ ☆ ☆ ᒧ
Grade 7	☆ ☆ ☆
Grade 8	☆ ☆ ☆ ☆ ☆

☆ = 50 hours

Grade 6: $4\frac{1}{2} \times 50 = 225$

Grade 7: $3 \times 50 = 150$

225 − 150 = 75

Grade 6 has _____ more hours than Grade 7.

Example 2

Use the pictograph above. How many fewer customers were there in May than in April?

There are $2\frac{1}{2}$ fewer symbols for May. Each symbol stands for 10 customers.

$2\frac{1}{2} \times 10 = 25$

There were 25 fewer customers in May than in April.

Practice 2

Use the pictograph above. How many fewer hours are there for Grade 5 than for Grade 4?

There are _____ fewer stars for Grade 5.

Each symbol stands for _____ hours.

_____ × _____ = _____

There were _____ fewer hours for Grade 5 than for Grade 4.

On Your Own

Use the pictograph above. In July there were 45 customers. How many symbols do you need to add July to the graph? _____

Write About It

Use the graph above. Write a new comparison you can make from the graph.

Objective 12.1: Represent and compare data by using pictures, bar graphs, tally charts, and picture graphs.

Lesson 12-1   **Compare Data with Graphs**

Try It

1. Use this graph.

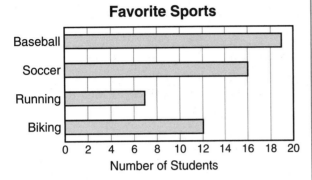

   a. How many more students chose soccer than biking? _____

   b. How many chose baseball? _____

2. Use this graph.

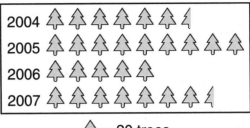

   🌲 = 20 trees

   a. How many more trees were planted in 2004 than in 2006? _____

   b. How many symbols would you need to show 70 trees? _____

3. Use the graph above. Which sport was chosen by the fewest students? Circle the letter of the correct answer.

   A baseball   B running

   C soccer   D biking

4. Use the graph above. The number of trees for 2003 was one-half that of 2006. How many trees were planted in 2003? Circle the letter of the correct answer.

   A 100   B 50   C 25   D $2\frac{1}{2}$

5. Use this graph.

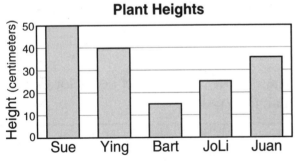

   Whose plants were more than twice as tall as Bart's plant?

   _____

6. Use the plant graph to the left. What is the range of plant heights? Show your work.

Objective 12.1: Represent and compare data by using pictures, bar graphs, tally charts, and picture graphs.

Lesson 12-2 — Record Data

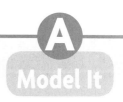

Activity 1

Ed used tally marks to record students' favorite colors. How many students did he ask?

Favorite Color

Red	Blue	Yellow	Green
𝍤𝍤	𝍤|	𝍤|||	𝍤𝍤||

The marks are in groups of 5.
Add the 5s, and then add the leftover marks.
There are 36 in all.

Practice 1

Kim took a color survey and got 9 votes for red, 7 for blue, 12 for yellow, and 15 for green.

Show Kim's results using tally marks.

Favorite Class

Red	Blue	Yellow	Green

How many people did she ask in all? _____

Activity 2

How many times does each letter of the alphabet occur in this passage?

> Cinco de Mayo is a celebration of the battle of Puebla, Mexico, on May 5, 1862. On that day a Mexican force defeated the French army.

Work with a partner. One partner reads each letter. The other partner uses tallies to record the frequency. Show the results in a chart.

a	b	c	d	e	f	g	h	i
12	3	7	4	14	5	0	4	5
j	k	l	m	n	o	p	q	r
0	0	3	5	6	9	1	0	4
s	t	u	v	w	x	y	z	
1	8	1	0	0	2	4	0	

Practice 2

How many times does each letter of the alphabet occur in this passage?

> The seven days of Kwanzaa are from December 26 to January 1. This holiday celebrates the heritage and culture of African Americans.

Work with a partner. Use tallies to record the frequencies. Then complete the chart below.

a	b	c	d	e	f	g	h	i
j	k	l	m	n	o	p	q	r
s	t	u	v	w	x	y	z	

On Your Own

Choose or write a passage with about 100 letters about your favorite holiday. Use tallies to find the frequencies of the letters.

Write About It

What 5 letters do you think are used most frequently? Give reasons for your answer.

Objective 12.2: Record numerical data in systematic ways, keeping track of what has been counted.

Lesson 12-2: Record Data

Words to Know   A **stem-and-leaf plot** organizes numbers by their tens digits.

Example

Make a stem-and-leaf plot to put these test scores in order.

78	82	75	65	82	87	62	95
87	82	65	78	98	95	80	92

Find the least number, 62. Use the tens digit for the stem and the ones digit for the leaf.

Key: 6|2 means 62

```
6 | 2
```

The greatest number is 98. So, use 6, 7, 8, and 9 for the stems. Put each number on the plot. Order the leaves from least to greatest.

```
6 | 2 5 5
7 | 5 8 8
8 | 0 2 2 2 7 7
9 | 2 5 5 8
```

Practice

Kamal is making robes for a singing group. He recorded the members' heights in centimeters.

Heights (Centimeters)

172	180	167	148	185	175	155	175
176	159	181	161	150	168	157	151
158	154	182	163	152	175	163	178

Show the heights in a stem-and-leaf plot.

Heights (Centimeters)

```
14 |
15 |
16 |
17 |
18 |
```

Key: 14|8 means 148

On Your Own

Agnes recorded the weights of the packages she mailed last week.

17	20	14	6	25	9	14	4	22

Show the weights on a stem-and-leaf plot.

Key: 1|4 means 14

Write About It

Use the test score stem-and-leaf plot in Example 1. Explain how to use the plot to find the range of the test scores. Show your work.

Objective 12.2: Record numerical data in systematic ways, keeping track of what has been counted.

Lesson 12-2 — Record Data

1. Isaac recorded the ages of the registered voters in his apartment building.

Ages (Years)

40	59	21	63	34	25	36
70	41	67	23	36	58	25
82	25	78	41	47	24	59

Show the ages in a stem-and-leaf plot.

Ages (Years)

2 |
3 |
4 |
5 |
6 |
7 |
8 |

Key: 3|4 means 34

2. Use the data in Exercise 1. Explain your answers.

  a. Is it easier to find the number of voters from the chart or the stem-and-leaf plot?

  b. Which data display is easier to use to find the youngest and oldest voters?

3. Nancy made this tally chart to show the colors of the bikes in the school bike rack. Circle the letter of the correct answer.

Bike Color

Black	White	Blue	Other
‖‖‖‖ ‖‖‖‖ ‖‖	‖‖‖‖ ‖‖‖	‖‖‖‖	‖‖‖‖ ‖‖‖‖ ‖

How many bikes did Nancy include in her color survey?

A 26   **B** 33   **C** 35   **D** 36

4. Ten students chose weather for their science projects, 9 chose electricity, 7 chose rocks, and 6 chose planets. Show the students' choices with tally marks.

Science Projects

Weather	Electricity	Rocks	Planets

5. Use the school bike data in Exercise 3.

  a. Show that the black bikes account for one-third of the total number.

  b. Which data display would be best to show the bike color data: bar graph, pictograph, or stem-and-leaf plot?

Objective 12.2: Record numerical data in systematic ways, keeping track of what has been counted.

Lesson 12-3: Mean, Median, and Mode

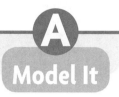

Words to Know — **Data** is information gathered by counting, measuring, asking, or observing. The **mean** of a set of data is the sum of the numbers divided by the number of numbers.

Activity 1

The cubes stand for the number of hours each student studied this week. Find the mean number of hours.

Mark	April	Maya	Tony	Chris

Rearrange to make the stacks equal.

When the stacks are equal, there are 3 in each stack. The mean is 3.

Practice 1

Corrine's restaurant sells fish for $11, grilled chicken for $8, waffles for $8, fruit salad for $6, veggie burgers for $7, and corn chowder for $8. What is the mean price of an item on her menu?

Use cubes to model the prices.

How many cubes are there? _____

How many stacks are there? _____

Rearrange to make the stacks equal.

How many cubes are in each stack? _____

Activity 2

Find the mean of this set: 3, 5, 5, 7.

Add all of the numbers. $3 + 5 + 5 + 7 = 20$

Divide the sum by the number of addends.
$20 \div 4 = 5$

The mean of the numbers is 5.

Practice 2

Find the mean of this set: 10, 8, 12, 11, 9.

Add. $10 + 8 + 12 + 11 + 9 =$ _____

Divide the sum by the number of addends.

_____ $\div$ _____ = _____

On Your Own

Neil's scores on his first four math tests were 74, 80, 74, and 92. What is the mean of his test scores?

Write About It

How does the mean of your scores represent your work in the class?

Objective 12.3: Find the mean, median, and mode of a data set.

Lesson 12-3: Mean, Median, and Mode

 Understand It

Words to Know The **median** is the middle value when data are listed from least to greatest. The **mode** is the number that occurs most often in a data set.

Example 1

Find the median of the data.
11, 16, 14, 2, 3, 2, 9

Order the data from least to greatest.
2, 2, 3, 9, 11, 14, 16

Cross out one number at a time on each end of the list until only one number is showing. This is the median.

2̶, 2̶, 3̶, 9, 1̶1̶, 1̶4̶, 1̶6̶

Practice 1

Find the median.
90, 44, 23, 83, 33, 110, 4, 71, 56

Order the data from least to greatest.

_____

Cross out one number at a time on each side of the list.

What is the median? _____

Example 2

Find the mode of these heights, given in feet.
65, 62, 63, 59, 71, 57, 62, 74, 58

Order the numbers from least to greatest.
57, 58, 59, 62, 62, 63, 65, 71, 74

Which number occurs most often? 62

The mode of the data is 62.

The most common height in this data set is 62 feet.

Practice 2

Find the mode of these ages, given in years.
11, 33, 45, 11, 23, 11, 61, 17, 23, 65

Order the numbers from least to greatest.

_____

What number occurs most often? _____

The mode of the data is _____.

The most common age in this data set is

_____.

On Your Own

Find the median and mode of the data.
31, 17, 18, 23, 15, 31, 19

Median: _____

Mode: _____

Write About It

Why is it helpful to put the numbers in a set of data in order when finding the mode of the data?

Objective 12.3: Find the mean, median, and mode of a data set.

Lesson 12-3 Mean, Median, and Mode

Try It

1. Find the median.

   a. 14, 12, 6, 4, 55 _____

   b. 17, 34, 29, 16, 34, 3, 20, 5, 32 _____

   c. 9, 90, 72, 42, 15, 38, 23, 102, 2, 7, 81 _____

2. Find the mean.

   a. 12, 12, 7, 9, 10 _____

   b. 3, 5, 5, 6, 3, 2, _____

   c. 31, 17, 18, 15, 19, 23, 31 _____

3. Find the mean, median, and mode of the data set: 30, 16, 17, 15, 18, 21, 30.

   Mean: _____

   Median: _____

   Mode: _____

4. What is the median amount that Sarah earned each week?

Sarah's Weekly Earnings				
Week 1	Week 2	Week 3	Week 4	Week 5
$20	$40	$25	$35	$50

   A $30   B $40

   C $35   D $25

5. Find the mean and median.

Play Tickets Sold					
Week	1	2	3	4	5
Tickets	150	170	220	160	220

   Mean: _____

   Median: _____

6. Karl's scores on the last 4 math tests were 100, 92, 65, and 63. What is the mean of these scores?

   _____

7. Pedro's scores on the last 4 math tests were 88, 90, 84, and 82. What is the mean of these scores?

   _____

8. Use your answers to Exercises 6 and 7. Which student's mean score is a better representation of his performance?

Objective 12.3: Find the mean, median, and mode of a data set.

Lesson 12-4   **Show Data in More Than One Way**

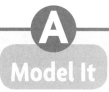

Words to Know   A **line plot** shows one X, or other symbol, for each data item.

Activity 1

Use the data in the chart to make a line plot.

Miles Run per Day

Sun	Mon	Tue	Wed	Thu	Fri	Sat
3	5	6	4	7	3	2

Write the days of the week below the line. Make an X for each mile run. Make 3 Xs for Sunday, 6 Xs for Tuesday, and so on.

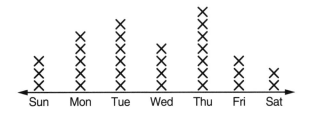

Practice 1

Use the data in the chart to make a line plot.

Movies Rented per Month

J	F	M	A	M	J	J	A	S	O	N	D
4	5	6	4	4	2	1	2	1	5	4	3

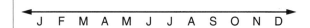

Activity 2

Use the data in the chart above to make a vertical bar graph.

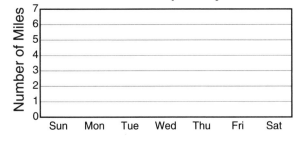

Practice 2

Use the data in the chart above to make a vertical bar graph.

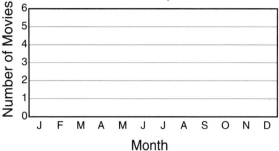
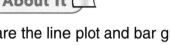

On Your Own

Dale has 4 siblings, Jenna has 2 siblings, Pedro has 1 sibling, and Marla has 3 siblings. Make a line plot to show this data.

Write About It

Compare the line plot and bar graph above. Which is easier to read? Which is easier to make? Give reasons for your answers.

Objective 12.4: Represent the same data set in more than one way.

Lesson 12-4 — Show Data in More Than One Way

B Understand It

Example 1

Use the data in the chart to make a pictograph. The key has been chosen for you.

School Club Membership

Chess	Drama	Math	Science

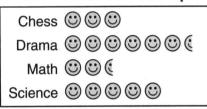

☺ = 10 students

Practice 1

Use the data in the chart to make a pictograph. The key has been chosen for you.

Weddings at City Hall

May	June	July	August
10	35	15	20

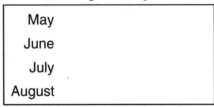

♡ = 5 weddings

Example 2

Use the data in the chart above to make a vertical bar graph.

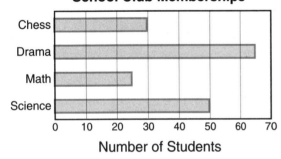

Practice 2

Use the data in the chart above to make a horizontal bar graph.

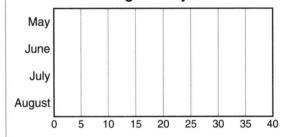

On Your Own

5 students had pizza for lunch, 3 had soup, 8 had sandwiches, and 5 had salads. Make a

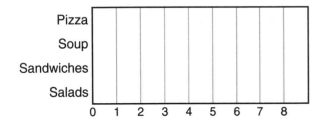

Write About It

Use the bar graph for the number of weddings at City Hall. Write labels for the horizontal and vertical axes. Why is it important to label graphs?

Objective 12.4: Represent the same data set in more than one way.

Lesson 12-4 — **Show Data in More Than One Way**

1. Use the data on the bar graph to complete the line plot.

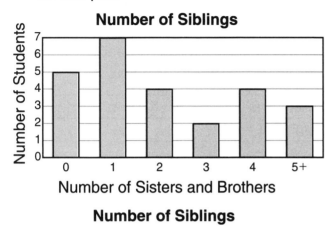

Number of Siblings

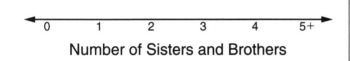

2. Use the data on the pictograph to complete the horizontal bar graph.

🏐 = 2 games

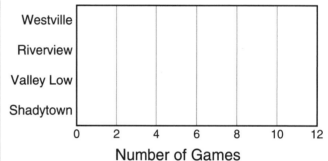

3. Use the data above. Which describes more than $\frac{1}{2}$ of the students? Circle the letter of the correct answer.

   A Have 3 or more siblings

   B Have 2 or more siblings

   C Have fewer than 2 siblings

   D Have no siblings

4. Use the data above to make a tally chart.

Objective 12.4: Represent the same data set in more than one way.

Lesson 12-5 | **Graph Ordered Pairs**

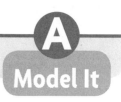

Activity

Su Li took a survey to find how many students in her school have pets. There were 25 third graders. Show this data on the grid below.

Start at 0. Go right to 3 for the third grade. Go up to 25, a point halfway between 20 and 30.

Draw the point.

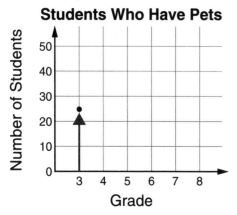

Practice

Raphael keeps track of the time he practices the tuba. Use the grid below to show his practice times in minutes for last week. The first day has been done for you.

Sun	Mon	Tue	Wed	Thu	Fri	Sat
30	45	30	60	60	15	75

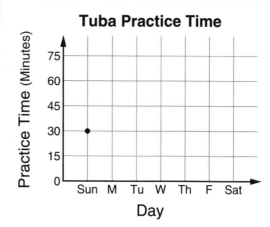

On Your Own

Use the data in this table to complete the graph above.

Students Who Have Pets

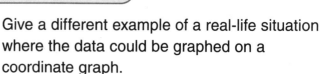

Grade	3	4	5	6	7	8
Number	25	40	35	20	45	30

Write About It

Give a different example of a real-life situation where the data could be graphed on a coordinate graph.

Objective 12.5: Use two-dimensional coordinate graphs to represent points and graph lines and simple figures.

Lesson 12-5: Graph Ordered Pairs

Words to Know The points at the corners of a geometric shape are called **vertices**. Each of these is a **vertex**.

Example

Graph these points.

a. Graph point D at (6, 5).

Start at 0. Go right to 6. Go up to the line that goes through 5. Write D to label the point.

b. Graph the triangle with vertices at A(2, 3), B(4, 1), and C(9, 2).

Graph each vertex with a point. Write the letters to label the points. Connect the points.

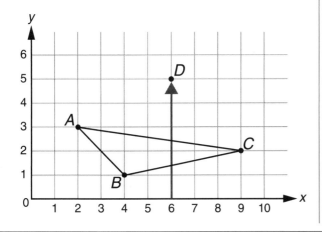

Practice

Graph these points.

a. point P at (2, 5)   b. point R at (9, 4)

c. point Q at (0, 4)   d. point S at (7, 0)

e. Graph the rectangle with vertices at W(2, 3), X(6, 5), Y(7, 3), and Z(3, 1).

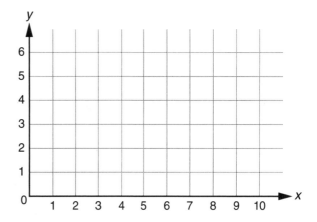

On Your Own

Identify a point on line segment AB. What are the coordinates of this point?

Write About It

What must be true about the coordinates of the endpoints of a vertical line segment? Include an example from the grid above.

Objective 12.5: Use two-dimensional coordinate graphs to represent points and graph lines and simple figures.

Lesson 12-5: Graph Ordered Pairs

Try It

1. Use the grid to graph the data in this table.

   Workers in Mr. Chen's Bike Shop

Sun	Mon	Tue	Wed	Thu	Fri	Sat
7	4	4	2	5	8	10

2. Use this grid.

   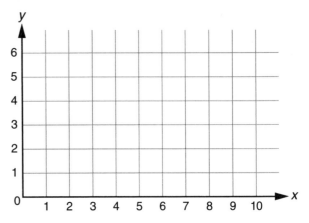

   Graph these points.

   a. point K at (2, 5)   b. point T at (4, 6)

   c. point M at (8, 0)   d. point Y at (6, 2)

   e. Graph the triangle with vertices at D(0, 2), E(3, 1), and F(6, 6).

3. Use the graph above. When does the shop have the greatest number of workers? Circle the letter of the correct answer.

   A Wednesday   B Friday

   C weekends   D weekdays

4. Use a first-quadrant grid. Graph these points: (1, 4), (3, 7), (6, 5), and (4, 2). Connect the points in order. Then connect the first and last points. What shape do you see?

   _____

5. Mark 3 points on a four-quadrant grid. Draw the triangle. Name the 3 points for your triangle.

6. Is the point for (3, 4) inside or outside the shape you drew in Exercise 4?

   _____

Objective 12.5: Use two-dimensional coordinate graphs to represent points and graph lines and simple figures.

Topic 12 — Graphing

Topic Summary

Choose the correct answer. Explain how you decided.

1. How many more bushels were there in 2006 than in 2004?

   A 300
   B 500
   C 600
   D 900

   Martinez Apple Orchards
   2004 🍎🍎🍎🍎🍎🍎
   2005 🍎🍎🍎🍎🍎🍎🍎🍎🍎
   2006 🍎🍎🍎🍎🍎🍎🍎
   2007 🍎🍎🍎🍎🍎

   🍎 = 200 bushels

___

2. When did the family travel the greatest distance?

   A During the 1st hour
   B During the 2nd hour
   C During the 6th hour
   D During the 8th hour

   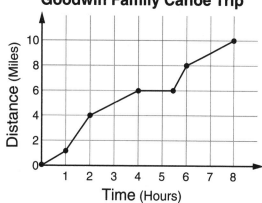
   Goodwin Family Canoe Trip

Objective: Review creating and interpreting graphs.

Topic 12 — Graphing

Mixed Review

1. Use the graph to the right. The populations have been rounded to the nearest 5,000.

   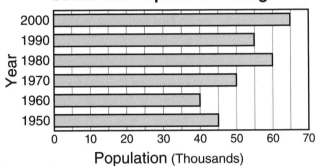
   Centerville Population Changes
   (Year vs. Population in Thousands)

   a. What was the population of Centerville in 1990? _____

   b. How did the population change from 1960 to 2000?
   _____

   Volume 5, Lesson 12-1

2. Compute.

   a. 824 + 236 = _____

   b. 21 + 431 = _____

   c. 28 + 39 = _____

   b. 509 + 74 = _____

   Volume 3, Lesson 8-3

3. What is the meaning of $n - 1$? Circle the letter of the correct answer.

   A One less than a number

   B One minus a number

   C A number subtracted from one

   D A number equal to one

   Volume 4, Lesson 11-2

4. Explain why a pictograph needs a key.

   Volume 5, Lesson 12-1

5. Write 6,702 in expanded notation.

   _____

   Volume 1, Lesson 3-2

6. When 4 people share 8 sandwiches equally, how much does each person get? What operation did you use to find out?

   Volume 2, Lesson 6-1

7. Estimate the quotient using compatible numbers: 614 ÷ 9.

   Volume 4, Lesson 10-2

Objective: Maintain concepts and skills.

Topic 13: Basic Geometric Figures

Topic Introduction

Complete with teacher help if needed.

1. Find the measure of the angle.

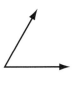

_____

Objective 13.1: Draw, measure, and classify different types of angles and lines.

2. Classify the polygon.

The polygon has _____ sides.

The polygon has _____ vertices.

The polygon is a _____.

Objective 13.2: Define polygon and classify the different types of polygons.

3. Classify the triangles.

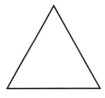

All sides are congruent.

It is a(n) _____ triangle.

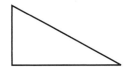

Exactly one angle measures 90°.

It is a(n) _____ triangle.

Objective 13.3: Explore, compare, and classify different types of triangles.

4. Identify the parts of the circle.

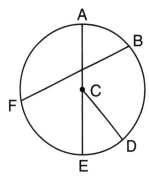

$\overline{AE}$ is the _____.

$\overline{BF}$ is a _____.

$\overline{CD}$, $\overline{AC}$, and $\overline{CE}$ are _____.

Objective 13.5: Explore circles and define their parts.

Lesson 13-1 Angles and Lines

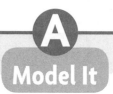
Model It

Words to Know

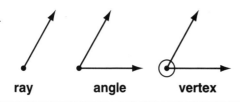
ray   angle   vertex

Activity 1

Measure the angle.

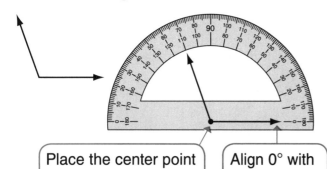

Place the center point of a protractor on the vertex of the angle.

Align 0° with one of the angle's rays.

Read where the other ray crosses the inside scale.

The angle measures 110°.

Practice 1

Measure the angle.

The angle measures _____.

Activity 2

Draw an angle whose measure is 35°.

Draw one ray. Align 0° on a protractor with that ray. Mark a dot at 35° on the inside scale. Connect the dot to the endpoint of the ray you drew.

Practice 2

Draw an angle whose measure is 180°.

On Your Own

Draw a 68° angle.

Write About It

What happens if you align one ray with 10 instead of 0?

Objective 13.1: Draw, measure, and classify different types of angles and lines.

Lesson 13-1 — Angles and Lines

Understand It — B

Words to Know

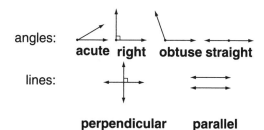

angles: acute, right, obtuse, straight

lines: perpendicular, parallel

Example 1

Classify each angle as acute, right, obtuse, or straight.

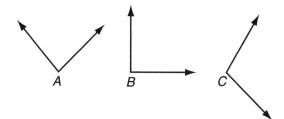

Measure each angle.
Angle A measures 85°, so it is **acute**.
Angle B measures 90°, so it is **right**.
Angle C measures 105°, so it is **obtuse**.

Practice 1

Measure. Classify each angle as acute, right, obtuse, or straight.

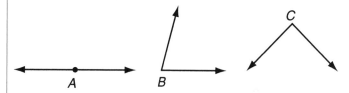

Angle A measures _____ so it is _____.

Angle B measures _____ so it is _____.

Angle C measures _____ so it is _____.

Example 2

Name a pair of parallel lines.

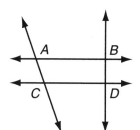

Lines $\overleftrightarrow{AB}$ and $\overleftrightarrow{CD}$ are the same distance apart at all points, so they are parallel.

Practice 2

Name a pair of perpendicular lines.

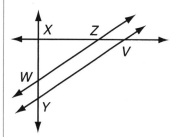

Lines _____ are perpendicular because they form _____ angles.

On Your Own

Draw a pair of lines that cross but are not perpendicular.

Write About It

Can you draw perpendicular lines that cross to form three 90° angles and one 20° angle?

Objective 13.1: Draw, measure and classify different types of angles and lines.

Lesson 13-1 **Angles and Lines**

1. Find the measure of the angle.

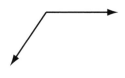

_____

2. Draw an angle with a measure of 70°.

3. Which pair of lines is parallel? Circle the letter of the correct answer.

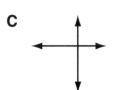

4. Which is an obtuse angle? Circle the letter of the correct answer.

5. What is the measure of the angle? Circle the letter of the correct answer.

A 45° **B** 50°

C 55° **D** 65°

6. What kind of angle is this? Circle the letter of the correct answer.

A acute **B** right

C obtuse **D** straight

7. Name a pair of perpendicular lines in the diagram below.

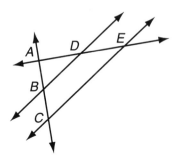

_____

8. Explain how to use a protractor to draw a 45° angle. Then draw the angle.

Objective 13.1: Draw, measure, and classify different types of angles and lines.

| Lesson 13-2 | Types of Polygons |

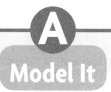

Words to Know A **polygon** is a closed figure made up of 3 or more **line segments** that meet but do not cross. The line segments are called **sides**. A **vertex** is the point where two sides of a polygon meet.

Activity 1

How many sides and vertices does a **quadrilateral** have?

Model a quadrilateral.

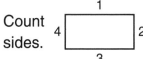

Count sides. Count vertices.

A quadrilateral has 4 sides and 4 vertices.

Practice 1

How many sides and vertices does a **triangle** have?

Model a triangle.

Count sides. Count vertices.

A triangle has _____ sides and _____ vertices.

Activity 2

How many sides and vertices does a **pentagon** have?

Model a pentagon.

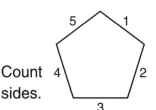

Count sides. Count vertices.

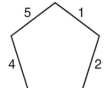

A pentagon has 5 sides and 5 vertices.

Practice 2

How many sides and vertices does an **octagon** have?

Model an octagon.

Count sides. Count vertices.

An octagon has _____ sides and _____ vertices.

On Your Own

How many sides and vertices does a hexagon have?

_____

Write About It

Explain why the figure below is not a pentagon.

Objective 13.2: Define *polygon* and classify the different types of polygons.

Lesson 13-2 — Types of Polygons

Words to Know
Tri- means 3.
Quad- means 4.
Pent- means 5.
Dec- means 10.
Hex- means 6.
Hept- means 7.
Oct- means 8.

Example 1

Classify the polygon.

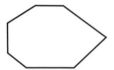

The polygon has 7 sides and 7 vertices.

The polygon is a **heptagon**.

Practice 1

Classify the polygon.

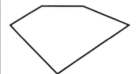

The polygon has _____ sides and _____ vertices.

The polygon is a _____.

Example 2

Classify the polygon.

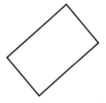

The polygon has 4 sides and 4 vertices.

The polygon is a **quadrilateral**.

Practice 2

Classify the polygon.

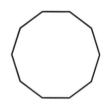

The polygon has _____ sides and _____ vertices.

The polygon is a _____.

On Your Own

Classify the polygon.

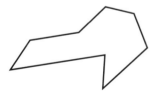

_____

Write About It

Explain why each figure below is not a polygon.

Objective 13.2: Define *polygon* and classify different types of polygons.

Lesson 13-2 — Types of Polygons

1. How many sides does each shape have?

   a. hexagon _____

   b. octagon _____

   c. pentagon _____

2. Draw two different quadrilaterals. Then circle the vertices.

3. Which figure is **not** a polygon? Circle the letter of the correct answer.

4. What is the name of the polygon? Circle the letter of the correct answer.

   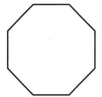

   A quadrilateral   B decagon

   C heptagon   D octagon

5. Which figure is **not** a quadrilateral?

6. Which figure is a heptagon?

        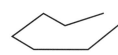

7. What polygon is used for each sign?

   _____ _____ _____

8. Answer *yes* or *no.* Explain.

   a. Is your math book a polygon?

   _____

   b. Is the shape of a page of your math book a polygon?

   _____

Objective 13.2: Define *polygon* and classify different types of polygons.

Lesson 13-3 — Triangles

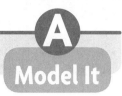
Model It

Words to Know — **Congruent** means having the same measure.
In **equilateral triangles**, all sides are congruent and all angles are congruent. In **isosceles triangles**, exactly two sides are congruent and exactly two angles are congruent.

Activity 1

Is the triangle **equilateral**?

Use a ruler. If all sides are equal, then the triangle is equilateral.

Each side is 2 cm, so all sides are equal.

The triangle is equilateral.

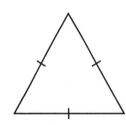

Practice 1

Circle the equilateral triangle.

Use a protractor. If all angles are equal, then the triangle is equilateral.

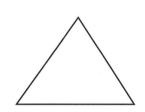 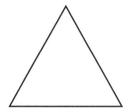

Activity 2

Is the triangle **isosceles**?

If exactly two sides are equal, then the triangle is isosceles. Use a ruler.

$\overleftrightarrow{AB}$ and $\overleftrightarrow{AC}$ are each 2 cm long.

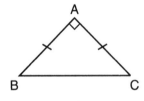

Practice 2

Circle the isosceles triangle.

Use a protractor. If exactly two angles are equal, then the triangle is isosceles. Mark the equal angles.

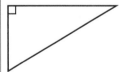

On Your Own

Classify the triangle.

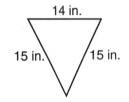

Write About It

Is it possible to draw an equilateral isosceles triangle? Explain why or why not.

Objective 13.3: Explore, compare, and classify different types of triangles.

Lesson 13-3 — Triangles

Words to Know   A **right triangle** has one right angle.

Example 1

The sum of a triangle's angle measures is 180°. Find the missing angle measure.

Add the angle measures: 60° + 60° = 120°.
Subtract from 180°: 180° − 120° = 60°.

The third angle measures 60°.

Practice 1

Find the missing angle measure. Show the operations.

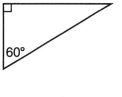

_____ + _____ = _____

_____ − _____ = _____

The third angle measures 30°

Example 2

Classify the triangle.

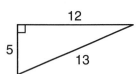

Examine the lengths of the sides. No sides are equal.

Examine the measures of the angles. There is 1 right angle.

The triangle is a **right triangle**.

Practice 2

Find the missing measures. Then circle the right triangle.

On Your Own

Find a triangle on this page that is both right **and** isosceles. Copy it here.

Write About It

Fill in the missing angle measure. Explain your answer.

In every equilateral triangle, each angle must measure _____° because

Objective 13.3: Explore, compare, and classify different types of triangles.

Lesson 13-3 **Triangles**

1. Classify the triangle.

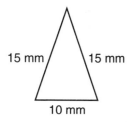
15 mm / 15 mm / 10 mm

_____

2. Classify the triangle.

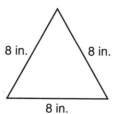
8 in. / 8 in. / 8 in.

_____

3. Which of the following best describes the triangle? Circe the letter of the correct answer.

A equilateral B isosceles

C right D right isosceles

4. Find the missing measure.

50°

5. Classify the triangle.

50° / 80° 50°

_____

6. Classify the triangle.

_____

7. Use a protractor. Label the angles. Then classify the triangle.

_____

8. Use a protractor. Label the angles. Then classify the triangle.

_____

Objective 13.3: Explore, compare, and classify different types of triangles.

Lesson 13-4: Quadrilaterals

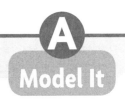
Model It

Words to Know A **quadrilateral** is a 4-sided polygon.
Congruent means having the same measure.

Activity 1

Find a polygon tile that is a rectangle.

Find the length of each side.
Opposite sides have equal lengths.

Find the measure of each angle.
Each angle measures 90°.

A **rectangle** has opposite congruent sides and four 90° angles.

Practice 1

Find a square, a trapezoid, and a rhombus. Which of these is also a rectangle?

Look for _____ sides with equal lengths.
Look for angles that measure _____.

The _____ is also a rectangle.

Activity 2

Find a polygon tile that is a parallelogram.

Opposite sides are parallel.

Find the length of each side.
Opposite sides have equal lengths.

Find the measure of each angle.
Opposite angles have equal measures.

A **parallelogram** has two pairs of parallel sides. Also, opposite sides are congruent, and opposite angles are congruent.

Practice 2

Find a rectangle, a rhombus, and a trapezoid. Which shapes are also parallelograms?

Look for sides that are _____.

The _____ are parallelograms.

On Your Own

Find the sum of the measures of the angles of any quadrilateral.

Write About It

Is a square a parallelogram? Explain.

Objective 13.4: Explore, compare, and classify different types of quadrilaterals.

Lesson 13-4 Quadrilaterals

B Understand It

Example 1

Classify the quadrilateral.

Examine the sides.
One pair is parallel.
No sides are congruent.

A quadrilateral with one pair of parallel sides is a **trapezoid**. The shapes below are also trapezoids.

Practice 1

Classify the quadrilateral.

Which sides are parallel? _____

Which sides are congruent? _____

Which angles are congruent? _____

The quadrilateral is a _____ and a _____.

Example 2

Find the measure of angle x.

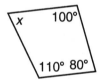

The sum of the angle measures of a quadrilateral is 360°.
$100° + 80° + 110° + x = 360°$
$290° + x = 360°$
$290° - 290° + x = 360° - 290°$
$x = 70°$
Angle x measures 70°.

Practice 2

Find the measure of angle x.

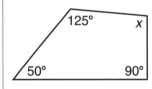

_____ + _____ + _____ + x = 360°

_____ + x = 360°

_____ − _____ + x = 360° − _____

x = _____

Angle x measures _____.

On Your Own

Complete the following sentence in as many ways as possible. A square is also a _____.

Write About It

Is the statement true? Explain.
All rhombuses are parallelograms.

Objective 13.4: Explore, compare, and classify different types of quadrilaterals.

Lesson 13-4 Quadrilaterals

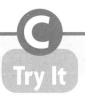

1. Classify the quadrilateral in as many ways as possible.

   _____

2. Classify the quadrilateral in as many ways as possible.

   _____
   _____

3. Which name **cannot** be used to describe the figure? Circle the letter of the correct answer.

   A quadrilateral   **B** square

   C trapezoid   **D** parallelogram

4. Which is **not** a quadrilateral? Circle the letter of the correct answer.

   A triangle   **B** rhombus

   C square   **D** trapezoid

5. A quadrilateral has 4 congruent angles and opposite sides that are parallel. Can you classify the figure as a square? Explain.

6. Find the measure of angle x.

   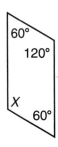

   _____

7. Draw a quadrilateral with 4 congruent sides but no right angles. What is the name of the quadrilateral you drew?

8. Is the statement true? Explain.
   All quadrilaterals are squares.

Objective 13.4: Explore, compare, and classify different types of quadrilaterals.

Lesson 13-5 Circles

Words to Know   A **circle** is a closed plane figure made up of points that are the same distance from the center. A **radius** is a line segment that connects the center of a circle to a point on the circle. The **diameter** is a line segment that connects two points on the circle and passes through the center.

Activity 1

Draw a circle with a radius of $\frac{3}{4}$ inch.

Use a compass. Draw a point that will be the center of your circle. Label the point.

Set the compass to the length of the radius. Hold the compass point at your center point. Move the compass to make a circle.

Practice 1

Draw a circle with a radius of $\frac{5}{8}$ inch.

Use a compass. Draw and label a point which will be the center of the circle.

Set the compass to _____ inch. Hold the compass point at the center point and move the compass to make a circle.

Activity 2

Draw a circle with a diameter of 1 inch.

Find the radius of the circle by dividing the diameter by 2.
1 inch ÷ 2 = $\frac{1}{2}$ inch

Use a compass to draw a circle with a radius of $\frac{1}{2}$ inch.

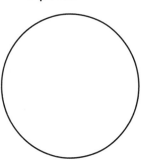

Practice 2

Draw a circle with a diameter of $\frac{1}{2}$ inch.

Find the radius of the circle.
$\frac{1}{2}$ inch ÷ 2 = _____

Use a compass to draw a circle with a radius of _____.

On Your Own

Draw a circle with a diameter of 2 inches.

Write About It

Explain how to find the diameter of a circle when you know the radius.

Objective 13.5: Explore circles and define their parts.

Lesson 13-5 Circles

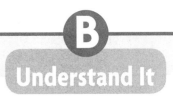

Words to Know A **chord** is a line segment that connects two points on a circle.

Example 1

Which segments are radii of circle C?

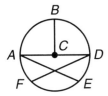

$\overline{AD}$, $\overline{AE}$, and $\overline{DF}$ are **chords** of circle C. Each connects two points on the circle.

$\overline{BC}$, $\overline{AC}$ and $\overline{DC}$ are **radii** of circle C. Each connects the center of a circle to one point on the circle.

Practice 1

Which segments are **chords** of circle B?

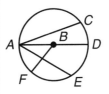

A chord connects two points on a circle.

_____, _____, and _____ are chords of circle B.

Example 2

Find the diameter of the circle.

The diameter of a circle is twice the radius.

Write an equation.
diameter = 2 × radius
diameter = 2 × 6.2
diameter = 12.4

The diameter of the circle is 12.4 inches.

Practice 2

Find the radius of the circle.

To find the radius of a circle, divide the diameter by _____.

Write an equation.
radius = diameter ÷ _____
radius = _____ ÷ _____
radius = _____

The radius of the circle is _____ mm

On Your Own

Which segments are diameters of circle A?

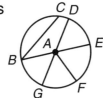

Write About It

Is every chord of a circle also a diameter of the circle? Explain.

Objective 13.5: Explore circles and define their parts.

Lesson 13-5 Circles

Try It

1. Draw a circle with a radius of $\frac{3}{8}$ inch.

2. Draw a circle with a diameter of $1\frac{1}{2}$ inches.

3. Which is a diameter of circle *A*? Circle the letter of the correct answer.

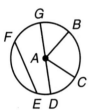

- **A** $\overline{EF}$
- **B** $\overline{DG}$
- **C** $\overline{AG}$
- **D** $\overline{AC}$

4. Which is **not** a chord of circle *C*? Circle the letter of the correct answer.

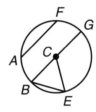

- **A** $\overline{AF}$
- **B** $\overline{BE}$
- **C** $\overline{BG}$
- **D** $\overline{CE}$

5. A pizza has a radius of 8 inches. What is the diameter of the pizza?

_____

6. A circular swimming pool has a diameter of 24 feet. What is the distance from the center of the pool to one edge?

_____

7. What is the diameter of the circle?

_____

8. What happens to the diameter of a circle if its radius is tripled?

Objective 13.5: Explore circles and define their parts.

Volume 5 — Level E

Topic 13: Basic Geometric Figures

Topic Summary

Choose the correct answer. Explain how you decided.

1. Which is an acute angle?

   A 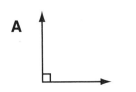 B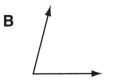

   C  D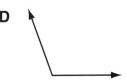

   _____
   _____
   _____

2. Which is **not** a quadrilateral?

   A rhombus

   B trapezoid

   C pentagon

   D parallelogram

   _____
   _____
   _____

Objective 13.5: Review basic geometric figures.

Topic 13: Basic Geometric Figures

Mixed Review

1. Élan ships gift packages to friends. He ships 9 packages. Each package costs $15 to ship. How much does Élan pay to ship his packages?

   _____

   Volume 4, Lesson 9-3

2. Which equation represents the following statement? Circle the letter of the correct answer.

   Eight more than some number is 13.

   A $n \div 8 = 13$ **B** $8 \div n = 13$

   C $13 + n = 8$ **D** $n + 8 = 13$

   Volume 4, Lesson 11-3

3. What is $78 \div 6$? Circle the letter of the correct answer.

   A 13 **B** 12

   C 14 **D** 15

   Volume 4, Lesson 10-3

4. Find the median and mode of the data.

   26, 29, 18, 31, 20, 29, 22

   Median: _____

   Mode: _____

   Volume 5, Lesson 12-3

5. Glenn worked 40 hours in 5 days. He worked the same schedule each day. How many hours did he work each day?

   _____

   Volume 2, Lesson 6-4

6. Brianna walked 6 miles on Saturday, 4 miles on Sunday, and 5 miles on Monday. How many miles did she walk all together? _____

   Volume 2, Lesson 4-4

Objective 13.5: Maintain concepts and skills.

Topic 14: Measurement Conversion

Topic Introduction

Complete with teacher help if needed.

1. Find each answer.

   a. To change from feet to inches, multiply by _____.

   b. To change from feet to yards, divide by _____.

   c. To change from centimeters to meters, divide by _____.

   d. To change from yards to inches, multiply by _____.

   Objective 14.5: Carry out simple unit conversions within a system of measurement.

2. Find each measure.

   a. 1 foot = _____ inches

   b. 1 yard = _____ feet

   c. 1 meter = _____ centimeters

   d. 1 yard = _____ inches

   Objective 14.5: Carry out simple unit conversions within a system of measurement.

3. Find each measure.

   a. 1 cup = _____ ounces

   b. 1 gallon = _____ quarts

   c. 1 quart = _____ pints

   d. 1 fluid ounce = _____ tablespoons

   Objective 14.5: Carry out simple unit conversions within a system of measurement.

4. Find each answer.

   a. 8 tablespoons = _____ fluid ounces

   b. 4 quarts = _____ pints

   c. 2 pints = _____ cups

   d. 2 gallons = _____ cups

   Objective 14.5: Carry out simple unit conversions within a system of measurement.

Lesson 14-1 U.S. Customary Units

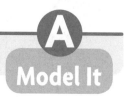

Words to Know

This table shows measures of **length**. Length is how long something is. Use a ruler to help you think about the measures.

inch	• about the length of half your thumb
foot	• 12 inches • a little longer than a piece of paper
yard	• 36 inches • 3 feet
mile	• 63,360 inches • 5,280 feet

This table shows measures of **weight**. Weight is how heavy something is.

ounce	• the weight of about 5 quarters
pound	• 16 ounces • about the weight of 3 baseballs
ton	• 2,000 pounds • about the weight of a car

Activity 1

What is the best unit for measuring:

a. the height of a house? feet or yards

b. the weight of a cell phone? ounces

Practice 1

What is the best unit for measuring:

a. the weight of a turkey? _____

b. the length of a worm? _____

Activity 2

Compare.

1 inch < 1 foot

2 pounds > 16 ounces

Practice 2

Compare. Write <, >, or =.

1 pound _____ 10 ounces

12 feet _____ 1 yard

On Your Own

What is the best unit for measuring:

a. the height of a giraffe? _____

b. the weight of a train? _____

Write About It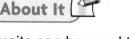

Which units can be used to measure the height of a door? Which is better? Explain.

Objective 14.1: Explore the basic units of measure in the United States.

Lesson 14-1 U.S. Customary Units

Words to Know The table shows measures for **capacity,** the amount a container holds.

fluid ounce	cup	pint	quart	gallon
• 2 tablespoons	• 8 fluid ounces	• 2 cups	• 4 cups • 2 pints	• 16 cups • 8 pints • 4 quarts

Example 1

What is the best unit for measuring:

a. milk in a pitcher? cups or pints

b. water for a hamster? fluid ounces

Practice 1

What is the best unit for measuring:

a. water in a swimming pool? _____

b. juice in a glass? _____

Example 2

Compare.

a. 1 pint < 1 quart

b. $\frac{1}{2}$ gallon > 1 pint

Practice 2

Compare. Write <, >, or =.

a. 2 pints _____ 1 quart

b. 1 gallon _____ 10 cups

On Your Own

Circle the better measure for the liquid in an eyedropper.

$\frac{1}{2}$ fluid ounce $\frac{1}{2}$ pint

Write About It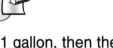

If 4 quarts are in 1 gallon, then there are 8 quarts in 2 gallons. How can you find the number of quarts in 3 gallons? Explain.

Objective 14.1: Explore the basic units of measure in the United States.

Lesson 14-1 — U.S. Customary Units

1. Which estimate of the weight of a pair of boots is best? Circle the letter of the correct answer.

 A about 2 ounces B about 2 gallons

 C about 2 pounds D about 2 tons

2. Which of the following is **not** true? Circle the letter of the correct answer.

 A 1 pint > 1 cup

 B 1 inch < 1 yard

 C 1 quart > 1 cup

 D 1 gallon < 1 fluid ounce

3. Compare. Write <, >, or =.

 4 pints _____ 1 gallon

 1 yard _____ 8 feet

4. Put in order from smallest to largest: pints, quarts, cups, gallons.

 _____

5. Which is **not** a unit for measuring capacity? Circle the letter of the correct answer.

 A ton   B quart

 C cup   D gallon

6. What is the best unit for measuring:

 a. the distance across an ocean? _____

 b. water in a wet kitchen sponge?

 _____

7. There are 2 pints in a quart. There are 4 quarts in a gallon. Finish the picture to show how many pints are in a gallon. Explain.

 | Pint |
 | Pint |

8. Write the most reasonable unit of length.

 a. The distance from New York to Los Angeles is 2,794 _____.

 b. The door of your classroom is about 2 _____ wide.

 c. The amount of weight a truck can pull is measured in _____.

Objective 14.1: Explore the basic units of measure in the United States.

Lesson 14-2 **Basic Metric Prefixes**

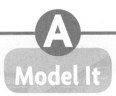
Model It

Words to Know   Milli- means one-thousandth. Centi- means one-hundredth.
Kilo- means one thousand.

1 meter = 1,000 millimeters	1 **meter** = 100 centimeters	1,000 meters = 1 kilometer
1 gram = 1,000 milligrams	1 **gram** = 100 centigrams	1,000 grams = 1 kilogram
1 liter = 1,000 milliliters	1 **liter** = 100 centiliters	1,000 liters = 1 kiloliter

Activity 1

Circle the smaller amount.

a. 1 kilogram   (1 gram)

b. 1,000 centimeters   (1 meter)

Activity 2

Circle the smaller amount.

a. 1,000 millimeters $\frac{1}{2}$ kilometer

b. 100 centimeters $\frac{1}{2}$ meter

On Your Own

Circle the smaller amount.

a. 10 kilograms   100 milligrams

b. 10 centimeters   1,000 millimeters

c. 100 milliliters   1 liter

Practice 1

Circle the smaller amount.

a. 1 millimeter   1 centimeter

b. 1 liter   100 milliliters

Practice 2

Circle the smaller amount.

a. 2 kilograms   200 grams

b. $\frac{1}{2}$ liter   10 milliliters

Write About It

How many millimeters are in 1 centimeter? Explain how you know.

Objective 14.2: Explore the basic metric prefixes and what they mean.

Lesson 14-2 | **Basic Metric Prefixes**

1 meter = 1,000 millimeters	1 **meter** = 100 centimeters	1,000 meters = 1 kilometer
1 gram = 1,000 milligrams	1 **gram** = 100 centigrams	1,000 grams = 1 kilogram
1 liter = 1,000 milliliters	1 **liter** = 100 centiliters	1,000 liters = 1 kiloliter

Example 1

Fill in the table to show equivalent amounts.

grams	1	2	3	4
centigrams	100	200	300	400

Practice 1

Fill in the table to show equivalent amounts.

liters	1	2	___
milliliters	1,000	___	3,000

Example 2

Fill in the table to show equivalent amounts.

meters	1,000	4,000	7,000	10,000
kilometers	1	4	7	10

Practice 2

Fill in the table to show equivalent amounts.

meters	1	2	___	49
centimeters	100	___	300	___

On Your Own

Fill in the blank.

400 centimeters = _____ meters

3 kilograms = _____ grams

Write About It

What is the difference between a milligram and a kilogram?

Objective 14.2: Explore the basic metric prefixes and what they mean.

Lesson 14-2: Basic Metric Prefixes

1. Fill in the blanks to show equivalent amounts.

meters	1	2	3	4
millimeters			3000	

2. 1,000 grams is equivalent to which of the following?

A 1 milligram
B 1 kilogram
C 100 milligrams
D 10 kilograms

3. Compare. Use <, >, or =.

a. 500 milliliters _____ $\frac{1}{2}$ liter

b. 500 centimeters _____ $\frac{1}{2}$ kilometer

4. Circle the smaller unit in each pair.

a. kilogram   milligram

b. meter   centimeter

5. Fill in the table to show equivalent amounts.

meters	1	10	100
centimeters	100		

6. Fill in the table to show equivalent amounts.

meters	1,000	3,000	5,000
kilometers	1		

7. Are any of these amounts equivalent? Explain.

100 millimeters, 10 centimeters, 1 meter

8. Eduardo is 2 meters tall. He wrote his height as 2,000 centimeters. Is this correct? Explain.

Objective 14.2: Explore the basic metric prefixes and what they mean.

Lesson 14-3: Use the Metric System

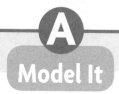
Model It

Words to Know

Length	
millimeter (mm)	• the height of a stack of 10 papers
centimeter (cm)	• the width of your little finger
meter (m)	• a little more than 3 feet

10 millimeters = 1 centimeter
1,000 millimeters = 1 meter
100 centimeters = 1 meter

Activity 1

Circle the best estimate for the depth of a diving pool.

3 centimeters   (3 meters)

Practice 1

Circle the best estimate for the height of a step.

15 centimeters   15 meters

Activity 2

Write <, >, or =.

5 millimeters < 1 centimeter

1 meter < 1,000 centimeters

Practice 2

Write <, >, or =.

3 centimeters _____ 6 millimeters

10 millimeters _____ 1 centimeter

On Your Own

Write the correct amounts to make the statements equal.

3 meters = _____ centimeters

$\frac{1}{2}$ meter = _____ millimeters

Write About It

A tree measures 2.5 meters tall. How many centimeters is this? Explain.

Objective 14.3: Explore the basic metric units and their relationships.

Lesson 14-3: Use the Metric System

Understand It — B

Words to Know — Mass is the amount of matter in an object.

Mass	
gram (g)	• a paper clip
kilogram (kg)	• equal to a little more than 2 pounds • 1,000 grams

Capacity	
milliliter (mL)	• about 20 drops of water
liter (L)	• a little more than 4 cups • 1,000 milliliters

Example 1

Circle the best estimate for the mass of a textbook.

2 grams   (**2 kilograms**)

Practice 1

Circle the best estimate for the amount of water an elephant might drink in one day.

100 milliliters   100 liters

Example 2

Write <, >, or =.

100 grams < 10 kilograms

1 liter > 100 milliliters

Practice 2

Write <, >, or =.

1,000 grams _____ 1 kilogram

2,000 milliliters _____ 1 liter

On Your Own

Write the correct amounts to make the statements equal.

4,000 grams = _____ kilograms

1 liter = _____ milliliters

Write About It

How many grams are in $\frac{1}{2}$ kilogram? Explain.

Objective 14.3: Explore the basic metric units and their relationships.

Lesson 14-3 | **Use the Metric System**

Try It

1. What is the best estimate for how much a glass of juice can hold?

   230 milliliters   23 liters

2. Is a millimeter, centimeter, or meter the best unit of measure for the height of a tree? Explain your reasoning.

3. Write <, >, or =.

   1.5 kilograms _____ 15 grams

   10 centimeters _____ 10 millimeters

4. Write <, >, or =.

   100 milliliters _____ 1 liter

   100 centimeters _____ 1 meter

5. Which is equivalent to 100 centimeters?

   A 0.1 meter   **B** 1 meter

   C 10 meters   **D** 100 meters

6. Which measure could be used to find the mass an apple?

   A gram   **B** liter

   C meter   **D** centimeter

7. Which of the following is **not** a measure of length?

   A millimeter   **B** milliliter

   C centimeter   **D** meter

8. How many grams are in 2 kilograms? Explain.

   _____

   _____

   _____

Objective 14.3: Explore the basic metric units and their relationships.

Lesson 14-4 **Factors in Unit Conversions**

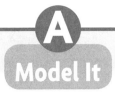

Words to Know Yards (yd), feet (ft), and inches (in.) are units of length in the U.S. system.

1 yard = 3 feet = 36 inches

1 foot = 12 inches

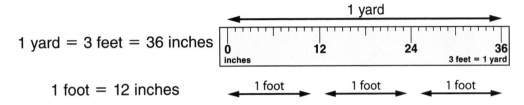

Meters (m) and **centimeters (cm)** are units of length in the metric system.

1 meter = 100 centimeters

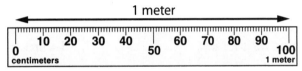

Activity 1

Write 1 foot, 2 feet, and 3 feet as inches. Each foot contains 12 inches, so multiply the number of feet by 12.

1 foot = 1 × 12 inches = 12 inches
2 feet = 2 × 12 inches = 24 inches
3 feet = 3 × 12 inches = 36 inches

Practice 1

Write 4 yards as inches. How many inches are in a yard? _____

Multiply the number of yards by _____.

4 yd = 4 × _____ in. = _____ in.

Activity 2

Write 7 meters as centimeters. Each meter contains 100 centimeters, so multiply the number of meters by 100.

7 m = 7 × 100 cm = 700 cm

Practice 2

Write 12 meters as centimeters. How many centimeters are in 1 meter? _____

12 meters = 12 × _____ cm = _____ cm

On Your Own

5 yd = (5 × _____) ft = _____ ft

15 ft = (15 × _____) in. = _____ in.

5 yd = _____ in.

Write About It

Explain why you multiply when changing from a larger unit to a smaller unit.

Objective 14.4: Express simple unit conversions in symbolic form.

Lesson 14-4 Factors in Unit Conversions

Words to Know To change from feet to yards, divide the number of feet by 3 since 3 feet = 1 yard.
To change from inches to yards, divide the number of inches by 36 since 36 inches = 1 yard.
To change from centimeters to meters, divide the number of centimeters by 100 since 100 centimeters = 1 meter.

Example 1

How many feet are in 48 inches?

Each foot contains 12 inches.
How many 12s are in 48?

Divide. 48 ÷ 12 = 4

48 inches = 4 feet

Practice 1

How many yards are in 18 feet?

1 yard = _____ feet

To find the number of yards in 18 feet, divide _____ by _____.

18 feet = _____ yards

Example 2

How many meters are in 800 centimeters?

Each meter contains 100 centimeters.
How many 100s are in 800?

Divide. 800 ÷ 100 = 8

800 centimeters = 8 meters

Practice 2

How many meters are in 1,400 centimeters?

1 meter = _____ centimeters

To find the number of meters, divide _____ by _____.

1,400 centimeters = _____ meters

On Your Own

288 in. = (288 ÷ _____) ft = _____ ft

24 ft = (24 ÷ _____) yd = _____ yd

288 in. = _____ yd

Write About It

Explain why you divide when changing from a smaller unit to a larger unit.

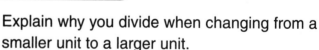

Objective 14.4: Express simple unit conversions in symbolic form.

Lesson 14-4: Factors in Unit Conversions

Try It

1. Fill in each blank.

   a. feet = yards × _____

   b. inches ÷ _____ = yards

   c. meters = centimeters ÷ _____

   d. feet × _____ = inches

   e. yards = feet ÷ _____

   f. inches ÷ _____ = feet

   g. centimeters = meters × _____

   h. inches = yards × _____

2. a. Find the number of inches in 5 feet.

   _____

   5 feet = _____ inches

   b. Find the number of meters in 700 centimeters.

   _____

   700 centimeters = _____ meters

3. Fill in each blank.

   a. To write yards as feet, multiply the number of yards by _____.

   b. To write feet as inches, multiply the number of feet by _____.

   c. To write inches as feet, divide the number of inches by _____.

4. Which statement is **not** correct? Circle the letter of the correct answer.

   A To change feet to inches, multiply the number of feet by 12.

   B To change centimeters to meters, divide the number of centimeters by 100.

   C To change feet to yards, multiply the number of feet by 3.

5. Jolene needs to add 3 yards, 7 feet, and 144 inches. She wants her answer to be a whole number using the largest unit possible. What unit should she use? Explain. What is the total length?

6. When you convert centimeters to meters, do you multiply or divide? Why?

Objective 14.4: Express simple unit conversions in symbolic form.

Lesson 14-5: Convert Units within a System

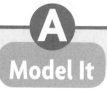
Model It

Words to Know A **measurement system** can be based on U.S. units or on metric units.

Activity 1

A fence is 4 yards and 3 feet long. How long is it in feet?

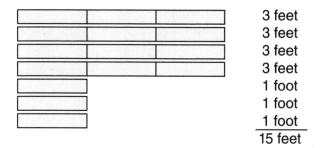

	3 feet
	3 feet
	3 feet
	3 feet
	1 foot
	1 foot
	1 foot
	15 feet

4 yards + 3 feet = 15 feet

Practice 1

Miko threw a ball 3 yards, 2 feet, and 17 inches. Convert the total distance to inches.

How many inches are in 3 yards? _____

How many inches are in 2 feet? _____

Add 17 inches.

What is the total number of inches that Miko threw the ball?

_____

Activity 2

The distance around a flower bed measures 6 meters and 185 centimeters. How long is it in centimeters?

1 meter	100 cm
1 meter	100 cm
1 meter	100 cm
1 meter	100 cm
1 meter	100 cm
1 meter	100 cm
	100 cm
	85 cm

6 meters + 185 centimeters = 785 cm

Practice 2

Ernestine used a measuring wheel to measure part of the length of a volleyball court. She rolled the wheel 14 meters and then rolled it 276 centimeters farther. Convert the total distance she measured to centimeters.

How many centimeters are in 14 m? _____

Add 276 centimeters.

What is the total number of centimeters Ernestine measured?

_____

On Your Own

Find the total number of centimeters in 7 meters and 152 centimeters.

_____

Write About It

Could you express 6 feet and 7 inches as a whole number of feet? Explain.

Objective 14.5: Carry out simple unit conversions within a system of measurement.

Lesson 14-5 | **Convert Units within a System**

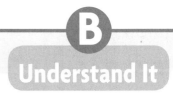

Words to Know Common units of time measurement are **seconds, minutes, hours,** and **days.** 60 seconds = 1 minute, 60 minutes = 1 hour, 24 hours = 1 day

Example 1

How many seconds are there in 1 hour, 10 minutes, 35 seconds?

  1 hour + 10 min + 35 sec

= (1 × 60 × 60) sec + (10 × 60) sec + 35 sec

= 3,600 sec + 600 sec + 35 sec

= 4,235 sec

Practice 1

How many seconds are there in 2 hours, 5 minutes, 8 seconds?

seconds in 2 hours = 2 × _____ × _____ = _____

seconds in 5 minutes = 5 × _____ = _____

Add 8 seconds.

total number of seconds = _____

Example 2

A bus trip took a total of 60 hours. How many days is this?

Each day has 24 hours.

24 + 24 = 48 hours = 2 days

60 hours − 48 hours = 12 hours or $\frac{1}{2}$ day

The trip took $2\frac{1}{2}$ days.

Practice 2

It takes $4\frac{1}{2}$ days to hike a mountain trail including time for rest breaks and meals. How many hours does it take to hike the trail?

How many hours are in each day?_____

How many hours are in 4 days? _____

How many hours are in $\frac{1}{2}$ day? _____

How many hours are in the hike? _____

On Your Own

Estelle left home at 2:00 P.M. She rode her bicycle for 75 minutes. At what time did she stop?

_____

Write About It

Shamir's goal is to practice the piano for 4 hours per week. On Monday he practiced 45 minutes, on Wednesday he practiced $1\frac{1}{2}$ hours, and on Saturday he practiced 95 minutes. Did he meet his goal? Explain.

Objective 14.5: Carry out simple unit conversions within a system of measurement.

Lesson 14-5 Convert Units within a System

Try It

1. Complete each of the following.

   a. 9 ft + 192 in. = _____ ft

   b. 9 ft + 192 in. = _____ in.

   c. 248 cm + 11 m + 52 cm = _____ m

   d. 248 cm + 11 m + 52 cm = _____ cm

2. Complete each of the following.

   a. 150 min + 7 hr + 120 min = _____ hr

   b. 150 min + 7 hr + 120 min = _____ min

   c. 18 hr + 5 days + 30 hr = _____ days

   d. 18 hr + 5 days + 30 hr = _____ hr

3. Complete each of the following.

   a. 5 hr + _____ min = $6\frac{1}{2}$ hr

   b. 13 ft + _____ in. = 18 ft

   c. 800 cm + _____ m = 12 m

   d. 2 days + _____ hr = $2\frac{1}{2}$ days

4. Which of the following is **more** than 120 sec + 3 min + 75 sec? Circle the letter of the correct answer

   A 350 sec   B 6 min

   C 7 min   D 200 sec

5. Eugene began his paper route at 5:30 A.M. If he must be done no later than 6:45 A.M. to catch the bus to school, what is the longest amount of time, in minutes, that his paper route can take?

   _____

6. John played soccer for 1 hour 30 minutes and jogged for 45 minutes. How long, in minutes, was his exercise time?

   _____

7. How many inches is 1 yard 1 foot 5 inches? Show your work.

8. Carmelita measured out three lengths of baseboard. The lengths of the baseboard pieces were 4 yards, 5 feet, and 72 inches. How many feet of baseboard did she have?

   _____

Objective 14.5: Carry out simple unit conversions within a system of measurement.

Topic 14 — Measurement Conversion

Topic Summary

Choose the correct answer. Explain how you decided.

1. There are 938 centimeters from Malcolm's front door to his mailbox. How many meters does Malcolm have to walk from his front door to his mailbox?

   A 9.38 m

   B 93.8 m

   C 9,380 m

   D 93,800 m

   _____

   _____

   _____

2. The softball team had throwing practice. One member of the team threw the softball 16 yards, 12 feet, and 84 inches. How many feet was the softball thrown?

   A 804 feet

   B 112 feet

   C 67 feet

   D 12 feet

   _____

   _____

   _____

Objective: Review measurement conversions.

Topic 14: Measurement Conversion

Mixed Review

1. Find each sum or difference mentally.

a. 8 + 7 = _____   b. 3 + 5 = _____

c. 20 − 6 = _____   d. 15 − 6 = _____

e. 12 − 7 = _____   f. 10 + 4 = _____

g. 9 + 8 = _____   h. 11 − 5 = _____

i. 17 − 10 = _____ j. 6 + 8 = _____

Volume 3, Lesson 8-1

2. Write each number in words.

a. 3,408

_____

_____

b. 4,809,126

_____

_____

_____

Volume 1, Lesson 3-3

3. Find each product mentally.

a. 6 × 4 = _____   b. 9 × 7 = _____

c. 3 × 5 = _____   d. 8 × 4 = _____

e. 2 × 7 = _____   f. 8 × 6 = _____

g. 9 × 3 = _____   h. 4 × 7 = _____

Volume 2, Lesson 5-5

4. For his trail mix, Devon bought 16 ounces of dried pineapples, 12 ounces of banana chips, 12 ounces of dried papaya, and 8 ounces of cashews. How many pounds of trail mix did he make?

Volume 5, Lesson 14-5

5. Complete the fact family.

30 ÷ 5 = 6

_____   _____

_____

Volume 2, Lesson 6-5

6. Which place is each digit in?

4,926

a. 4 _____

b. 6 _____

c. 2 _____

d. 9 _____

Volume 1, Lesson 3-1

Objective: Maintain concepts and skills.

Topic 15: Measure Geometric Figures

Topic Introduction

Complete with teacher help if needed.

1.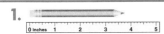

a. Between which two inch markings is the length of the pencil?

_____

b. Is the length of the pencil more or less than $3\frac{1}{2}$ in.?

_____

c. Is the length closer to 3 in. or 4 in.?

_____

Objective 15.1: Measure the length of an object to the nearest inch or centimeter.

2.

a. How many squares are in each row?

_____

b. How many rows are there?

_____

c. How many squares are there all together?

_____

Objective 15.3: Estimate or determine the area of figures by covering them with squares.

3.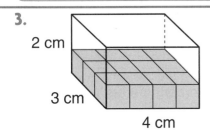

a. How many cubes are in the bottom layer?

_____

b. How many layers will fit in the box?

_____

c. How many cubes will fill the box?

_____

Objective 15.5: Estimate or determine the volume of solid figures by counting the number of cubes that would fill them.

4.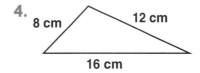

a. What are the lengths of the 3 sides of the triangle?

_____

b. What is the total distance around the triangle?

_____

Objective 15.2: Find the perimeter of a polygon with integer sides.

Lesson 15-1: Length

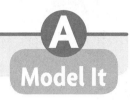

Words to Know
An **inch** is a unit of length in the U.S. system.
A **centimeter** is a unit of length in the metric system.

Activity 1

When measuring to the nearest inch, determine which inch marking on the ruler the length is closest to. Begin a measurement from the 0-inch position on the ruler.

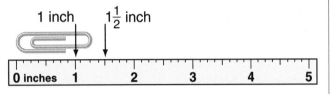

The paper clip is a bit longer than 1 inch but is less than $1\frac{1}{2}$ inches. To the nearest inch, the paper clip is 1 inch long.

Practice 1

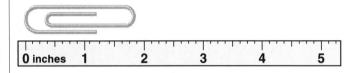

Is the paper clip longer than 1 inch? _____

Is it shorter than 2 inches? _____

Which inch mark is closest? _____

Activity 2

The pinto bean below measures more than 1 centimeter but less than the 2 centimeters. It is closer to 2 centimeters. To the nearest centimeter, the bean is 2 centimeters long.

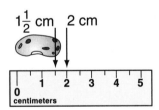

Practice 2

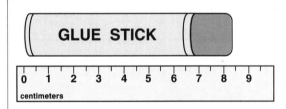

Between which two centimeters is the end of the glue stick?

_____

Which is it closest to? _____

On Your Own

Draw a line 1 inch long. Give the length of your line to the nearest centimeter.

_____

Write About It

Would you measure very short distances to the nearest inch or nearest centimeter? Explain.

Objective 15.1: Measure the length of an object to the nearest inch or centimeter.

Lesson 15-1: Length — Understand It

Example 1

Measure this pencil. To the nearest inch, what is its length?

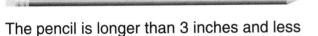

The pencil is longer than 3 inches and less than halfway between 3 and 4 inches.

To the nearest inch, the pencil is 3 inches long.

Practice 1

Between which two inch-lengths is the length of the notepad? _____

To the nearest inch, what is the length of the notepad? _____

Example 2

Measure this highligher. To the nearest centimeter, what is its length?

HIGHLIGHTER

The highlighter is longer than 6 cm and less than halfway between 6 and 7 cm.

To the nearest centimeter, the highlighter is 6 cm long.

Practice 2

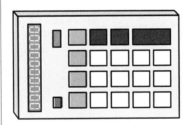

To the nearest centimeter, what is the length of the calculator? _____

On Your Own

Measure the length of your math book to the nearest inch and the nearest centimeter. Which number is greater? Why?

_____

_____

Write About It

Jesse owns the pencil in Example 1. Which pencil case should he buy, one that is 3 inches long or one that is 4 inches long? Does finding the nearest inch help you? Explain.

Objective 15.1: Measure the length of an object to the nearest inch or centimeter.

Lesson 15-1 **Length**

Try It

1. Give the length of each arrow to the nearest inch.

   a.  _____

   b. 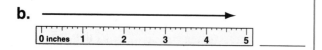 _____

   c. 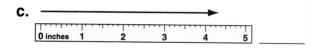 _____

2. Give the length of each arrow to the nearest centimeter.

   a.  _____

   b.  _____

   c.  _____

3. Draw an arrow for each measurement.

   a. 3 inches to the nearest inch

   b. 5 centimeters to the nearest centimeter

4. What is the length of this rectangle to the nearest centimeter?

5. Find three things in your classroom that are 12 inches to the nearest inch.

6. Look at the rectangle above in Exercise 4. What is its length to the nearest inch?

Objective 15.1: Measure the length of an object to the nearest inch or centimeter.

Lesson 15-2 Perimeter

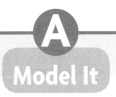

Words to Know A **polygon** is a two-dimensional shape that is closed and has segments for all of its sides. The **perimeter** of a polygon is the distance around all its sides.

Activity 1

The perimeter is found by counting all the edges of 1-cm squares you pass as you move around the figure as shown by the arrows.

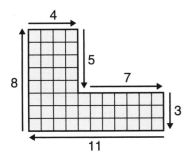

Perimeter = (4 + 5 + 7 + 3 + 11 + 8) cm
= 38 cm

Practice 1

Each side of each square is equal to one inch. Give the length of each side in inches.

A to B _____

B to C _____

C to D _____

D to E _____

E to F _____   Perimeter = _____

F to A _____

Activity 2

Find the perimeter of a triangle in which all sides measure 8 ft.

Number of sides in a triangle = 3

Each side measurement = 8 ft

Perimeter = 3 × 8 ft = 24 ft

Practice 2

Find the perimeter of a square in which all sides measure 13 m.

Number of sides in a square = _____

Each side measurement = _____

Perimeter = _____

On Your Own

The perimeter of a triangle is 37 inches. One side measures 15 inches, and a second side measures 12 inches. What does the third side measure?

Write About It

How would finding the distance around a circle be different from finding the distance around a rectangle? Explain.

Objective 15.2: Find the perimeter of a polygon with integer sides.

Lesson 15-2 Perimeter

B Understand It

Words to Know A **regular polygon** is one whose side measurements are all the same.
A **pentagon** is a polygon with 5 sides.
A **hexagon** is a polygon with 6 sides.
An **octagon** is a polygon with 8 sides.

Example 1

Find the perimeter of a rectangle with a length of 12 feet and a width of 7 feet.

A rectangle has 2 lengths and 2 widths.

length = 12 ft
width = 7 ft   width = 7 ft
length = 12 ft

Perimeter = [(2 × length) + (2 × width)] ft
= [(2 × 12) + (2 × 7)] ft
= [24 + 14] ft
= 38 ft

Practice 1

Find the perimeter of a rectangle with a length of 17 m and a width of 12 m. Label the sides of this rectangle with the appropriate measurements.

Perimeter = [(2 × _____) + (2 × _____)] m

= [_____ + _____] m = _____ m

Example 2

Find the perimeter of a regular octagon whose sides measure 16 in.

Perimeter = 8 sides × 16 in. per side = 128 in.

Practice 2

Find the perimeter of a regular pentagon whose sides measure 14 cm.

_____

On Your Own

A regular hexagon and a regular octagon each have a perimeter of 24 inches. What is the measurement of each side of each polygon?

_____

_____

_____

Write About It

If the perimeter of a square is 16 yd and you increase its perimeter to 24 yd, by how much does that change each side? Explain.

Objective 15.2: Find the perimeter of a polygon with integer sides.

Lesson 15-2 **Perimeter**

1. Find the perimeter of each figure.

   a. _____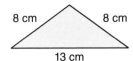

   b. _____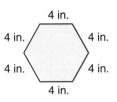

   c. _____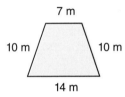

2. Determine the perimeter of each described polygon.

   a. a rectangle with a length of 6 ft and a width of 4 ft

   _____

   b. a triangle with sides of length 14 cm, 19 cm, and 23 cm

   _____

   c. a regular pentagon with sides of length 7 in.

   _____

3. Answer each of the following for a rectangle with a length of 8 in. and a width of 5 in.

   a. What is its perimeter? _____

   b. What would its perimeter be if the length is doubled? _____

4. What is the perimeter of a rectangle with a length of 9 ft and a width of 7 ft? Circle the letter of the correct answer.

   A 16 ft   B 32 in.

   C 32 ft   D 63 ft

5. A planting bed is in the shape of a regular hexagon. Each side measures 12 feet. What is the distance you go if you walk all the way around the planting bed?

   _____

6. A rectangular garden is 18 feet long and 13 feet wide. How much fencing is required to surround the garden?

   _____

7. What is the perimeter of a regular triangle with sides of length 6 in.?

   _____

8. What is the perimeter of a square with sides of length 12 feet?

   _____

Objective 15.2: Find the perimeter of a polygon with integer sides.

Lesson 15-3 Area

Model It

Words to Know   **Area** is the amount of surface the shape has. Area is measured in **square units,** such as square inches (sq in.) or square centimeters (sq cm).

Activity 1

What is the area of the figure below?

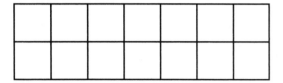

Count the squares to find the area. The area of the rectangle is 14 square centimeters.

Practice 1

What is the area of the figure below?

Count the squares to find the area.

_____

Activity 2

The rectangle is 4 inches long and 3 inches wide. Its area in square inches is 12 sq in.

4 sq in. per row
× 3 rows
———
12 sq in.

Practice 2

Draw a rectangle that is 6 inches long and 5 inches wide. Draw lines to show how to divide the rectangle into 1-inch squares. What is the area of the rectangle?

_____

On Your Own

Trace your hand, with your fingers together, on a sheet of centimeter grid paper. Give the approximate area of your hand in square centimeters.

_____

Write About It

If you are using graph paper to find the area of a circle and the area of a square, which measurement would be more accurate? Explain.

Objective 15.3: Estimate or determine the area of figures by covering them with squares.

Lesson 15-3 Area

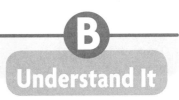

Example 1

This box is 3 cm long, 2 cm wide, and 1 cm high. What is the total area of all six sides?

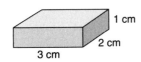

Front + back = ▢ + ▢ = 6 sq cm
Left + right = ▢ + ▢ = 4 sq cm
Top + bottom = ▢ + ▢ = 12 sq cm

Total area: (6 + 4 + 12) sq cm = 22 sq cm

Practice 1

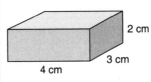

Use grid paper to draw each side of the box. Count the squares to find the total area of all sides.

Front + back = _____

Left + right = _____

Top + bottom = _____

Total area: _____

Example 2

This circle is covered with 1-cm squares. One estimate of its area might be 16 cm². Another estimate may be slightly more because there are parts of the circle not covered by the squares.

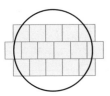

Practice 2

The triangle to the right has been covered with 1-inch squares.

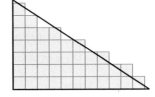

Estimate the area of the triangle. _____

On Your Own

Use centimeter grid paper. Draw a circle, a rectangle, and a triangle on the grid paper. Count squares to estimate the area of each figure in square centimeters.

_____

_____

Write About It

A rectangle has an area of 18 square inches. What is one possibility for the length and width of this rectangle? Explain.

Objective 15.3: Estimate or determine the area of figures by covering them with squares.

Lesson 15-3   **Area**

1. The grid paper shows square inches. Estimate the area of each figure.

   a. _____

   b. _____

   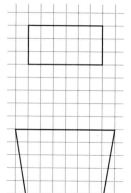

2. The grid paper shows square inches. Estimate the area of each figure.

   a. _____

   b. _____

3. What is the area of the rectangle shown below? Circle the letter of the correct answer.

   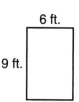

   A 15 feet   **B** 15 square feet

   C 54 feet   **D** 54 square feet

4. What is the minimum amount of paper needed to cover this cube? Circle the letter of the correct answer.

   A 384 cm   **B** 512 cm^3

   C 384 cm^2 **D** 512 cm^2

5. What is the total area of each of these boxes?

   a.

   _____

   b.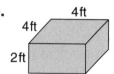

   _____

6. What is the area of the figure shown below?

   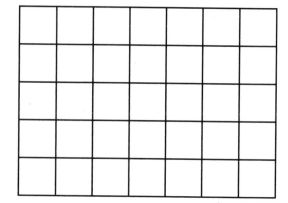

   _____

Objective 15.3: Estimate or determine the area of figures by covering them with squares.

Lesson 15-4: Area of Rectangles

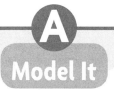

Words to Know The **area** of a figure is the number of square units needed to cover its surface. A **square unit** is a unit of area with dimensions 1 unit by 1 unit.

Activity 1

You can find the area of the rectangle using the formula Area = Length × Width.

Area = 8 × 2 = 16

The area of the rectangle is 16 square centimeters or 16 cm².

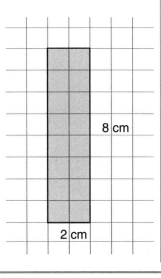

Practice 1

Draw a rectangle on centimeter grid paper and find the area in square centimeters.

What is the area of your rectangle?

_____

Activity 2

Enrique measures the area of the cover of his math book. He uses an inch ruler to measure the length and width.

Area = Length × Width

Area = 11 in. × 9 in. = 99 in²

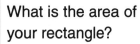
11 in.

9 in.

Practice 2

Find the area of your math book's cover and the area of your desktop. Choose two different units of measure. Show your work.

On Your Own

Find the area of a rectangle with length of 12 feet and a width of 6 feet. Show your work.

_____

_____

Write About It

What units of measure did you choose for the area of the book cover and the area of the desktop in Practice 2? Explain why you choose them.

Objective 15.4: Measure the area of rectangular shapes by using appropriate units.

Lesson 15-4: Area of Rectangles

Understand It — B

Example 1

What would be appropriate units of measure to describe the area of the state of Colorado?

A state has a very large area. You would want to use a large unit.

Either square miles or square kilometers would be an appropriate unit of measure.

Practice 1

What would be appropriate units of measure for the area of a playground? Explain.

_____

_____

_____

_____

Example 2

What is the area of the tablecloth?

Area = Length × Width

Area = 54 × 45 = 2,430 in^2

The area of the tablecloth is 2,430 in^2.

45 in. | tablecloth
54 in.

Practice 2

A painting is in a case measuring 135 cm by 60 cm. What is the area of the front of the case?

60 cm
135 cm

On Your Own

Damon has a chicken pen in the shape of a rectangle. It is 12 feet wide and 15 feet long. What is the area of the pen?

_____

Write About It

Would it be more appropriate to measure the area of your classroom floor in square centimeters or square meters? Explain.

Objective 15.4: Measure the area of rectangular shapes by using appropriate units.

Lesson 15-4 Area of Rectangles

Try It

1. What does each square on the grid represent? Find the area of the figure.

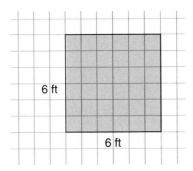

Each square represents _____.

Area = _____

2. Choose an appropriate unit of measure for each area.

For **a** and **b**, write in², ft², or mi².

a. basketball court _____

b. a greeting card _____

For **c** and **d**, write cm², m², or km².

c. a country _____

d. a postcard _____

3. Find the area of each rectangle.

a. length = 9 ft, width = 2 ft _____

b. length = 17 cm, width = 5 cm _____

c. length = 4 yd, width = 13 yd _____

d. length = 125 m, width = 6 m _____

4. Measure the figure. What is its area? Circle the letter of the correct answer.

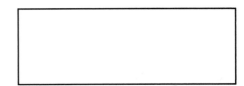

A 8 cm   B 12 cm²

C 12 m²   D 16 cm

5. A tabletop has an area of 28 square feet. Could a tablecloth that is 48 inches by 60 inches cover the tabletop? Explain.

6. What is the total area of the figure?

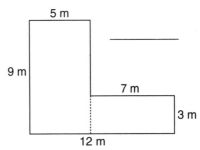

7. What is the area of a rectangle with a length of 10 in. and a width of 4 in.? Circle the letter of the correct answer.

A 14 in²   B 40 in²

C 40 in   D 28 in

8. Fill in the formula and find the area for this rectangle. Be sure to include units.

25 cm
6 cm

_____ × _____ = _____

Objective 15.4: Measure the area of rectangular shapes by using appropriate units.

Lesson 15-5 Volume

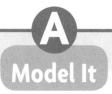

Words to Know The **volume** is the number of cubic units needed to fill the space occupied by a solid or three-dimensional figure.
Cubic units are used to measure volume and tell the number of cubes of a given size that are needed to fill a three-dimensional figure.
A **rectangular solid** is a three-dimensional figure with 6 rectangular faces.

Activity 1

Use centimeter cubes to build a rectangular solid with length = 2 cm, width = 2 cm, and height = 3 cm.

Create the first layer:

How many cubes are in the first layer?
4 cubes

Practice 1

Use centimeter cubes to build a rectangular solid. Choose a length, width, and height.

L = _____ cm, W = _____ cm,

H = _____ cm

Create the first layer. How many cubes are in the first layer? _____

Activity 2

Complete the figure above.

How many layers of cubes are in the completed figure? 3 layers

How many cubes are in the completed figure? 12 cubes

Each cube is 1 cubic centimeter or 1 cm³, so the volume of the rectangular solid is 3 layers cubic centimeters or 12 cm³.

Practice 2

Complete the figure above.

How many layers of cubes are in the completed figure? _____

How many cubes are in the completed figure?

_____

What is the volume of your rectangular solid?

_____

On Your Own

Use 24 cubes to create a rectangular solid. What are the dimensions of the solid?

_____

Write About It

When you know the number of cubes in one layer and the number of layers, how can you find the total number of cubes?

Objective 15.5: Estimate or determine the volume of solid figures by counting the number of cubes that would fill them.

Lesson 15-5 Volume

B Understand It

Example 1

The rectangular solid is made of inch cubes.

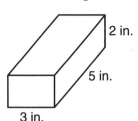

How many layers of cubes are there? 2
How many cubes in each layer? 5 × 5 = 15
How many cubes make up this figure? 30
What is the volume? 30 in³

Practice 1

The rectangular solid is made of centimeter cubes.

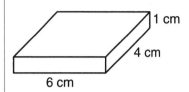

How many cubes make up this figure? _____

What is the volume of the rectangular solid?

_____

Example 2

A rectangular solid contains 5 layers of centimeter cubes. The first layer is shown below.

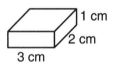

How many cubes are in each layer of the figure? 6

How many layers are in the figure? 5

What is the total number of cubes in the figure? 6 × 5 = 30

What is the volume of the rectangular solid? 30 cm³

Practice 2

A rectangular solid contains 10 layers of inch cubes. The first layer is shown below.

How many cubes are in each layer of this figure? _____

How many layers are in the figure? _____

What is the total number of cubes in the figure? _____

What is the volume of the rectangular solid?

_____

On Your Own

Construct a rectangular solid with a volume of 15 cubic centimeters. What are the dimensions of the solid?

_____

Write About It

If the dimensions of a rectangular solid are given in meters, what unit would you use to report the volume? Why?

Objective 15.5: Estimate or determine the volume of solid figures by counting the number of cubes that would fill them.

Lesson 15-5 — **Volume**

1. The figure is made from centimeter cubes. What is the volume of the figure?

   _____

2. The figure is made from centimeter cubes. What is the volume of the figure?

   _____

3. What are the dimensions of the rectangular solid in Exercise 1?

   _____

4. What are the dimensions of the rectangular solid in Exercise 2?

   _____

5. A rectangular box can be filled with 7 layers of cubic inches. Each layer contains 10 cubic inches. What is the volume of the box? Circle the letter of the correct answer.

   A 10 in³   **B** 17 in³

   C 35 in³   **D** 70 in³

6. Box A and Box B have the same length and width, but different heights. A layer of 8 cubic centimeters fits perfectly in the bottom of each box. Box A can hold 3 of these layers; Box B can hold 6 of these layers.

   a. What is the volume of Box A? _____

   b. What is the volume of Box B? _____

   c. What is the difference between the volumes of Box A and Box B?

   _____

7. Construct a rectangular solid with a volume of 18 cubic centimeters. What are the dimensions of the solid?

   _____

8. If the number of layers making up a rectangular solid is doubled but the other dimensions stay the same, what happens to the solid's volume? What happens if the number of layers is tripled? Explain.

Objective 15.5: Estimate or determine the volume of solid figures by counting the number of cubes that would fill them.

Topic 15: Measure Geometric Figures

Topic Summary

Choose the correct answer. Explain how you decided.

1. Find the area of a rectangle with a length of 12 inches and a width of 8 inches.

   A 96 inches

   B 40 inches

   C 96 square inches

   D 20 square inches

   _____
   _____
   _____

2. Find the perimeter of a rectangle with a length of 14 feet and a width of 9 feet.

   A 126 square feet

   B 126 feet

   C 46 feet

   D 23 feet

   _____
   _____
   _____

Objective: Review measuring geometric figures.

Topic 15 — Measure Geometric Figures

Mixed Review

1. Perform each operation.

   a. 78 + 89 = _____

   b. 95 − 37 = _____

2. Convert each measurement to the given unit.

   a. 36 in. = _____ ft

   b. 400 cm = _____ m

   c. 8 yd = _____ in.

   d. 7 m = _____ cm

   Volume 3, Lesson 7-2 and 7-3 — *Volume 5, Lesson 14-5*

3. What is 614 − 529? Circle the letter of the correct answer.

   A 85   B 115

   C 185   D 95

4. Which rounds **up** to 500? Circle the letter of the correct answer.

   A 523   B 440

   C 580   D 462

   Volume 3, Lesson 8-3 — *Volume 1, Lesson 3-4*

5. Write each number in expanded notation.

   a. 852 _____

   b. 8,607 _____

6. Use the commutative and associative properties of multiplication.

   a. If 11 × 13 = 143, then 13 × 11 = _____.

   b. If 4 × (8 × 3) = 96, then (4 × 8) × 3 = _____.

   Volume 1, Lesson 2-2 — *Volume 2, Lesson 5-3*

Objective: Maintain concepts and skills.

Words to Know/Glossary

A

acute angle — An angle that measures less than 90 degrees.

angle — Two rays or segments that meet at a common endpoint.

area — Area is the amount of surface the shape has. It is measured in square units.

C

capacity — The measure of how much something holds.

centi- — A metric prefix meaning one-hundredth.

centimeters — A centimeter (cm) is a measuring unit of length in the metric system. 1 meter = 100 centimeters.

chord — A line segment that connects two points on a circle.

circle — A closed plane figure made up of points that are the same distance from the center.

circumference — The distance around a circle.

congruent — Having the same measure.

cubic units — Units used to measure volume and tell the number of cubes of a given size that are needed to fill a three-dimensional figure.

D

data — Information collected about people or things.

days — A common unit of time measurement.

diameter — A chord that passes through the center of the circle.

E

equilateral triangle — A triangle with all sides congruent and all angles congruent.

F

feet — A unit of length in the U.S. system. 1 foot (ft) = 12 inches.

G

gram — A metric unit for measuring mass.

H

height — The height of a figure is the length of a perpendicular line between the base and the top of the figure.

hexagon — A polygon with 6 sides.

horizontal axis — The horizontal axis on a graph is at the bottom. The numbers go from left to right.

hours — A common unit of time measurement. 24 hours = 1 day.

I

inches — An inch (in.) is a unit of length measurement in the U.S. system. 36 inches = 3 feet = 1 yard.

isosceles triangle — A triangle with exactly two sides congruent and exactly 2 angles congruent.

K

key — The key on a pictograph shows the value of each symbol.

kilo- — Metric prefix meaning one thousand.

kilogram — A kilogram (kg) is metric unit for measuring mass. 1 kg = 1,000 g.

L

length — The measure of how long something is.

line — A set of points that continues without end in both directions.

line plot — A line plot shows one X, or other symbol, for each data item.

line segment — A part of a line between two endpoints.

liter — A metric unit for measuring capacity.

M

mass — The amount of matter in an object.

mean — The sum of the addends in a set of data divided by the number of addends.

measurement system — A measurement system can be based on U.S. units or on metric units.

median — The middle number in an ordered set of data.

meter — A metric unit for measuring length or distance.

mile — A unit of length measurement in the U.S. system. 1 mile = 63,360 inches = 5,280 feet.

milli- — A metric prefix meaning one-thousandth.

milliliter — A metric unit for measuring capacity; 1,000 mL = 1 L.

millimeter — A millimeter (mm) is a measuring unit of length in the metric system. 1 meter = 1,000 millimeters.

minutes — A common unit of time measurement. 60 minutes = 1 hour.

mode — The number that occurs most often in a set of data.

O

obtuse angle — An angle that measures greater than 90 degrees but less than 180 degrees.

octagon — A polygon with 8 sides.

ordered pair — An ordered pair shows the location of a point. The first number is the distance from the x-axis, and the second number is the distance from the y-axis.

ounce — A U.S. unit for measuring weight.

P

parallel lines — Lines that lie in the same plane and do not intersect.

pentagon — A polygon with 5 sides.

perimeter — The distance around all sides of a figure.

perpendicular lines — Lines that intersect to form right triangles.

pi (π) — A ratio of circumference of a circle to its diameter; it is approximately 3.14.

pictograph — A graph that uses symbols or simple pictures to represent quantities.

polygon — A closed figure made up of three or more line segments that meet but do not cross.

pound — A U.S. unit for measuring weight; 16 ounces.

Q

quadrilateral — A 4-sided polygon.

R

radius — A line segment that connects the center of a circle to a point on the circle.

range — The difference between the greatest and least numbers in a set of data.

ray — Part of a line that has one endpoint and continues without end in one direction.

rectangular solid — A three-dimensional figure with 6 rectangular faces.

regular polygon — One whose side measurements are all the same.

right angle — An angle that measures greater than 90 degrees.

right triangle — A triangle with one right angle.

S

seconds — A common unit of time measurement. 60 seconds = 1 minute.

sides — The line segments that make up a polygon.

square units — A unit of area with dimensions 1 unit by 1 unit.

stem-and-leaf plot — A stem-and-leaf plot organizes numbers by their tens digits.

straight angle — An angle measuring 180 degrees.

T

ton — A U.S. unit for measuring weight; 2,000 pounds.

V

vertex (pl: vertices) — The point where two sides of a ray or a polygon meet.

vertical axis — The vertical axis is at the left. The numbers increase from bottom to top.

volume — The number of cubic units needed to fill the space occupied by a solid or three-dimensional figure.

W

weight — The measure of how heavy something is.

X

x-axis — On a coordinate grid, the *x*-axis is horizontal.

Y

yards — A yard (yd) is a unit of length. There are 36 inches in 1 yard.

y-axis — On a coordinate grid, the *y*-axis is vertical.

Word Bank

Word	My Definition	My Notes

Word	My Definition	My Notes

Index

A
acute angle, 21
angles, 20–22
area, 62–64
    of rectangles, 65–67

C
capacity, 39
centigram, 42
centimeter, 41–44, 47, 56
chord, 33
circles, 32–34
comparing data with graphs, 2–4
congruent, 29
congruent triangles, 26
cubic units, 68
cup, 39

D
data
    comparing with graphs, 2–4
    definition of, 8
    recording, 5–7
    showing in more than one way, 11–13
day, 51
dec-, 24
diameter, 32
drawing angles and lines, 20–22

E
equilateral triangles, 26

F
fluid ounce, 39
foot, 38, 47

G
gallon, 39
gram, 41, 46
graphs
    comparing data with, 2–4
    graphing ordered pairs, 14–16

H
hept-, 24
heptagon, 24
hex-, 24
hexagon, 23, 60
hour, 51

I
inch, 38, 47, 56
isosceles triangle, 26

K
key, 3
kilometer, 42–43

L
length, 38, 44, 56–58
line plot, 11
line segment, 23
lines, 20–22
liter, 41

M
mass, 46
mean, 8
mean, median, and mode, 8–10
measurement system, 50
median, 9
meter, 41–47
metric system
    metric prefixes, 41–43
    metric units, 44–46
mile, 38
milliliter, 42
millimeter, 43–44
minute, 51
Mixed Review
    12: Graphing, 18
    13: Basic Geometric Figures, 36
    14: Measurement Conversion, 54
    15: Measure Geometric Figures, 84
mode, 9

O
oct-, 24
octagon, 23, 60
ordered pairs, graphing, 14–16
ounce, 38

P
parallelogram, 29
pent-, 24
pentagon, 23, 60
perimeter, 59–61
pictograph, 3
pint, 39
polygons, 23–25, 59–61
pound, 38

Q
quad-, 24
quadrilaterals, 23–24, 29–31
quart, 39

R
radius, 32–33
range, 2
recording data, 5–7
rectangles, 29
    area of, 65–67
rectangular solid, 68
regular polygon, 60
right angle, 21
right triangle, 27

S
second, 51
sides, 23
square units, 62
stem-and-leaf plot, 6

T
ton, 38
Topic Summary
    12: Graphing, 17
    13: Basic Geometric Figures, 35
    14: Measurement Conversion, 53
    15: Measure Geometric Figures, 83
trapezoid, 30
tri-, 24
triangles, 23, 26–28

U
unit conversion
    factors and divisors in, 47–49
    within a system, 50–52
units of U.S. measure, 38–40

V
vertex, 15, 23
volume, 68–70

Y
yard, 38, 47

W
weight, 38